The Dimensions of Federalism

D1367555

For my family

Contents

Tables

Preface to the Paperback Edition

A cartoon on a recent editorial page illustrates one of the ongoing debates over the role of state governments in America. A large block of concrete hangs menacingly over the heads of several figures drawn to represent the poor, sick, and elderly in society. In large letters on the hovering object is the term "Block Grants." Obviously, the cartoon reflects an oft-heard argument that if state governments are given unrestricted policy responsibilities, certain groups will likely suffer.

Arguments about the impact of state governments on public policy are hardly new. Dire warnings of drastic state revisions to national policies date back to James Madison's writings. Glowing promises of innovation and flexibility at the state level have roots in de Tocqueville. Yet, fundamental questions remain open to debate. When programs are delegated to state governments, what is the likely impact on policy? What kinds of leadership will state policy-makers provide? Will the strongest programs develop where most needed? Will innovations exceed federal standards? Will effective programs be emulated by other states?

These questions occupy a central position in much of the current dialogue concerning federalism and public policy. Even before 1994, the political climate in the United States fostered many proposals and promises for decentralization of policymaking through delegation to state and local levels of government. While not always consistent in its actions, the Reagan administration did make the transfer of authority to the state level a common theme of its domestic policy announcements. Presidents Bush and Clinton also, to varying degrees, voiced their support for replacing the federal presence in many policy areas with greater state autonomy and discretion. The most dramatic advocacy of decentralization occurred, however, after the Republican congressional victories of 1994. Republican majorities in the 104th Congress proposed replacing long-standing categorical grant programs with block grants, supported restrictions on unfunded federal mandates, and generally promised a massive devolution of power from federal to state and local governments. Advocates of these proposals promised that the changes would free states to deliver goods and services more effectively and efficiently. Congressional Democrats, liberals, a variety of policy analysts, and even some editorial cartoonists expressed concerns over proposed changes by arguing that potential state leadership would be undercut by irresponsibility, incompetence, or fiscal stress. Indeed, the answers to the questions raised above suggest the widest range of possible outcomes.

Those questions about state leadership also provide the central focus of this book. I offer answers to each based on a fairly simple theoretical model and both quantitative and qualitative analysis. The empirical examination considers state government efforts in pollution control policies following the substantial restructuring of federal and state roles in the early 1970s. I argue that state leadership in each policy area is shaped by the horizontal dimension of state competition and the vertical dimension of federal influence. While the empirical analysis thus focuses on specific policies, I contend that the argument is applicable to other policy areas at other points in time that also involve potential competition between states and some degree of federal influence. This includes many of the issues that received so much attention in 1995 such as welfare and health care as well as the very same environmental policies discussed in this book.

Indeed, the environmental policies discussed in this book remain essential components of governmental redress of pervasive problems. Air quality improved in some urban areas during the early 1990s, notably in Los Angeles where the program described in chapter 4 was implemented and then strengthened in 1994. Still, stationary source and mobile source air pollution continue to plague many urban areas. The EPA estimates that, in 1993, 25 million Americans lived in counties that violated the federal standard for fine particulates alone. Further, ozone and the acid rain problems (described in chapter 2) violate geopolitical boundaries to affect visibility and health even in such seemingly pristine places as Shenandoah and Grand Canyon national parks.

Water pollution from point and nonpoint sources continues to prevent attainment of water quality goals. Contamination occasionally has caused major public health crises such as the 1993 cryptosporidium outbreak in Milwaukee that killed over 100 people and sickened thousands of others. Millions of people still receive water from systems that violate EPA standards. Even with innovative programs such as those in North Carolina (chapter 3) and Iowa (chapter 5), nearly half of American rivers and lakes still do not meet standards for designated usage by 1994.

This paperback edition is thus identical to the hardback version published in 1992. The questions about state leadership have not gone away. Indeed, they are even more pervasive in today's political climate. I think that my answers to these questions continue to be pertinent and important for environmental policies, federalism, and the political system in general.

Acknowledgments

I believe in the old saying that the journey makes the destination worthwhile. Affirmation of such wisdom at the point of reaching this destination, the publishing of this book, may seem strange, but it is not intended to demean the final product. Rather, I think the journey to this point is worthy of acknowledgment. In this case, the project has involved several years, many locations, and numerous individuals whose contributions warrant more than these few words. Nevertheless, I will do my best.

I dedicate this book to my family. My father, B.J., and Tom have provided wisdom, patience, and love that kept me going even when I was tempted to unplug the computer and go clean outhouses in Yosemite for a living. I've been told that I inherited the will to write from my mother. I thank her and offer the hope that what follows does not make it too apparent that skills are not easily transferred across generations.

This work began as a dissertation at Stanford University. There I was blessed with some great friends, some helpful advisers, and four years of fellowship money. My committee of John Ferejohn, Gary Jacobson, Terry Moe, and Roger Noll pushed and challenged me until I was determined to do better than just getting by. I think you will like this version better than the last one. I was fortunate to receive help and advice from many people: John Bendor, Bill Bianco, Dorothy Blake, Dick Brody, John Bruni, Bill Caler, Dale Cochran, Leslie Eliason, David Fallek, Chris Gacek, Emily Goldman, Bob Griffin, John Hannah, Scott Johnson, Jack Levy, David Lumsdaine, Lee Metcalf, Kathy Moffeit, Rick Olguin, Harry Paposotierieu, Ken Roberts, Larry Rothenberg, Chris Scholl, Lewis Shepherd, Chuck Shipan, Steve Stedman, Serge Taylor, Kathy Teghtsoonian, George Tsebelis, Mike Webb, and Mark Woodward. The Harnishes provided an always appreciated refuge from academia. Finally, without the encouragement of Judy Goldstein and her family, this thing would be a prospectus propping up a table leg some-place.

Thanks to the Brookings Institution, I finished the dissertation and started the book while enjoying a year in Washington, D.C. Between softball games, the project was shaped by comments from Sarah Bales, Maureen Casama-you, John Chubb, Bob Copeland, Carol Evans, John Hird, Bob Katzmann, Pam Lokken, Tom Mann, Phil Mundo, Pietro Nivola, Paul Peterson, Paul Pierson, Mark Rom, Robin Smith, Steve Smith, Andy Sobel, Susan Stewart, Kent Weaver, and Joe White. The Opekas provided a place to live and a constant reality check.

I finished the book while employed by Washington University. I learned and continue to learn from my colleagues and my students. Those contribut-ing directly to this project include Barry Ames, Debbie Beckmann, Charles Franklin, Chris Gilbert, John Gilmour, Nancy Green, Larry Heimann, John Kautsky, Steve Kay, Ann Kendrick, Jack Knight, Liane Kosaki, Carol Mer-shon, Bob Salisbury, Julie Shermak, John Sprague, Paul Wahlbeck, Brian Werner, Julie Withers, and the Political Science Department staff. Again, I am blessed with family nearby and thank the Bersells, Blacks, and Browns for offering home away from home.

Since this is a book about state governments, I would be remiss if I did not thank those who have taught me much about this subject, those who work there. I will mention just a few: Dan Johnston, James Rickun, and Dale Ziege in Wisconsin; Ken Eagleson, Gary Hunt, Allan Klimek, and Don Safrit in North Carolina; Eugene Fisher and Tom Cackette in California; Ubbo Agena, Jim Gulliford, Darrell McAllister, Bill and Cathy Pielsticker, and Jerry Shepter in Iowa. Other officials, both at the state and federal level, are hereby thanked anonymously.

Finally, I want to acknowledge Larry Malley, the Duke staff, and two anonymous reviewers who proved to me that this process really can work. The book is better due to their efforts. If it still does not satisfy some, the fault lies with the author. After all, I'm the only one who has been on this journey since the start. Besides, Yosemite would not be a bad place to end up.

1 A Theory of State Leadership

On 10 August 1989 officials of eight northeastern states announced new regulations on automobile emissions that were stricter than those proposed by President Bush just weeks earlier. Observers and journalists expressed some surprise that this initiative by the states exceeded the policy response that the national government had formulated. They should not have been surprised.

Maligned and criticized for two centuries, state governments have arguably been undergoing a transformation in recent years. Criticisms of state parochialism date to Madison's description of subnational variations on national policy as the principal impediment to effective American government.[1] Since then, state governments have been characterized as backwaters for the worst excesses of American politicians and bastions for the most shameful of American policies.[2] Fear and mistrust of state governments greatly contributed to the centralized nature and hierarchical structure of domestic policies that now shape American political life. Much of this criticism was justified, but an alternative perspective is currently gaining support.[3] Some now argue that some state governments, as the result of a recent revitalization, have taken the lead in policy efforts in the United States. This perspective asserts that state governments, with some qualifications, "now are arguably the most responsive, innovative, and effective level of government in the American federal system."[4]

How can we analyze this asserted leading role for state governments in the federal system? We need to assess both the consistency and the exceptionalism of state leadership. How widespread are appropriate responses to policy needs among state governments? What makes the exceptional programs outstanding? What impact do they have? To understand the leadership of state governments, in this book I consider overall state behavior and the actions of leading state programs.

Consideration of the former, the overall behavior of state governments, describes the variance in state responsiveness and enables identification of state leaders in specific policy areas. If we define policy needs as the relative lack of publicly demanded goods, the most apparent evidence of responsive state behavior is the ability to match local policy supplied to local policy needed. State governments could provide national leadership through consistently appropriate responsiveness to societal needs. This is indeed the promise of a federal system. However, even those who espouse the state resurgence argument do not suggest that this promise has yet been realized.[5] The transformation of state governments has not occurred across the board. As a result, variance in the performance of state governments is evident in most policies. Thus I offer comparative analysis of all fifty states to discuss the extent of responsive state behavior and to identify state leaders as those state governments with the most developed programs in specific areas of policy needs.

Analysis of the latter phenomenon, the efforts of those state leaders, generates answers to basic questions of leadership. What do state leaders do? How do they disseminate ideas? How do they affect developments in other states? For states to take the lead in policy responsiveness, they must have followers. After all, leadership involves not only creativity and innovation, but also communication and organization. Therefore, I examine the states that have developed responsive programs and discuss the depth and impact of their developments.

My basic thesis is that state leadership, including both the overall matching of response to need and the influence of specific programs, is affected by intergovernmental dimensions of the policy involved. These dimensions, to be more fully discussed shortly, concern the vertical involvement of the federal government in state behavior and the horizontal potential for interstate competition.

The context of this analysis is the behavior of states and their leadership role in the policy areas of pollution control. Pollution control involves many specific issues. This analysis is directed at four of them: stationary source air pollution, point source water pollution, mobile source air pollution, and nonpoint source water pollution. These four areas are discussed for both pragmatic and theoretical reasons. Practically, they comprise some of the most important and consequential aspects of environmental protection in the United States. Theoretically, they constitute four cases covering a range of situations that vary along dimensions important in a federal system of government.

Why is analysis of this thesis in this context worthy of consideration? The efficacy of pollution control policies, as I show in subsequent chapters, depends heavily on state behavior. Since pollution control is important to quality of life, the prevalence of responsive behavior and the efforts of leading states may hold important implications. State policymakers can lead by formulating appropriate programs, innovating, exceeding minimum standards, disseminating ideas, and coordinating other state efforts. If state leadership is determined by the federal dimensions of horizontal competition and vertical involvement, then the ultimate development and efficacy of those pollution control policies will vary according to those same dimensions.

If my arguments are supported, then states do not consistently shirk their policy responsibilities nor do they always provide the momentum behind effective policy response. This seemingly simple point contradicts the perception by the public that irresponsible state pandering to industry has stimulated many aspects of pollution control statutes; the characterizations of state governments that have shaped interest group behavior on these issues; and the analyses that have blamed states for numerous policy failures in these areas. It is also counter to excessively positive descriptions of state behavior used to justify political demands for the decentralization of domestic policies gaining prominence in recent years. If the dimensions of federalism do have an impact on state behavior, then the analysis has important implications for many American policy efforts. The role of state governments in the initiation and implementation of numerous domestic programs may warrant reconsideration. Reconsideration will benefit from analysis that accounts for the effects of structural dimensions of federalism.

This chapter is designed to accomplish several things. First, this introductory section has identified the leadership of state governments as the dependent variable, the phenomenon to be explained. Next, I utilize different perspectives on federalism to build a model of state behavior. The extant literature suggests that overall state behavior is variable and subject to numerous influences besides policy needs. Third, I place the model within the federal context by describing dimensions for differentiating the federal policies that states must implement. Horizontal competition and vertical involvement are defined as the dimensions of federalism. The completed model is then used to generate hypotheses about the extent, depth, and impact of state leadership in specific policies. Fourth, I provide some background for the following chapters by discussing pollution control prior to the significant federal interventions of the 1970s and the subsequent differentiation of policies. Finally, I discuss the design of the remainder of the study.

A Model of State Government Behavior

States matter. Policies are not simply created by national officials and then routinely implemented by state and local governments as if they were unquestioning automatons in some Weberian machine. Rather, state officials make policy and adjust national efforts to match parochial circumstances. Here lies both the danger and the promise of federalism. The danger of a federal system is that subnational policymakers will respond only to private, not public, demands, thereby skewing policies to the extent that outcomes no longer match national intentions. The promise of a federal system is that subnational autonomy will provide flexibility, innovation, and efficiency by allowing policymakers who are close to the scene to tailor policy efforts to local public needs. For state governments to be leading policy efforts, state responses must be appropriate.

State responses to policy needs have not displayed the extremes in performance that would validate either the worst fears or the highest hopes of a federal system. State policies are neither totally dominated by internal economic considerations nor completely responsive to public needs. A model of state government behavior must therefore include factors suggested by several perspectives. This section reviews those perspectives.

The general argument motivating this review is that no single factor dominates state behavior. If any one factor alone did determine state behavior, then further examination of the leadership role of state governments would be unnecessary. For example, if the severity of the problem at hand solely determined state responses, then all one would need to know to understand state behavior would be the level of policy need. Responsiveness would be constant and the potential impacts of leading state programs would not matter. Dissemination of state innovations across state lines, coordination of state efforts, and the willingness of state officials to exceed federal guidelines would be inconsequential. Thus, perfect matching of state response to state need is offered as the fallible argument, or "straw man," that is to be countered by other perspectives.

Hypothesized matching of response and need

Political analysts have long extolled the virtues of a system of government that enables close contact between the people and the policymakers. Montesquieu praised small republics wherein "the interest of the public is more obvious, better understood, and more within the reach of every citizen."[6] A system of federalism wherein different levels of government share responsibilities can therefore not only provide buffers to dangerous concentrations of power, but can also facilitate citizen participation and involvement in important decisions at subnational levels. As a famous exponent argues, the "virtue of the federal system lies in its ability to develop and maintain mechanisms vital to the perpetuation of the unique combination of governmental strength, political flexibility, and individual liberty, which has been the central concern of American politics."[7] Hypothetically, subnational policies can and will be directed at the problems at hand by interested, involved citizens.

This classical perspective has been the subject of renewed interest in recent years. In the late 1980s, following a period during which subnational levels were criticized for seeming to be unresponsive, some observors described a revitalization of state governments that promised an important role for the influence of policy needs on state behavior. State governments were characterized as building responsible and professional political institutions that could provide effective policy responsiveness to current problems.[8] These authors described state governments as potentially "the new heroes of American federalism, implementing national policies throughout their borders in a responsible and responsive manner."[9] For intergovernmental programs, this argument suggested significant potential for responsible and professional implementation of domestic policies. Several reasons underlie this perspective.

First, representation has become more equitable. Due to civil rights legislation and judicial decisions such as *Baker v. Carr* (one person, one vote), states now benefit from ideas and perspectives from a variety of previously unheard viewpoints. Considerable evidence for this development exists. For example, between 1970 and 1988, the number of black state legislators doubled and the percentage of women holding state office quadrupled.[10] These developments do not by themselves guarantee higher quality legislation at the state level, but they do suggest increased potential for fresh ideas and approaches.

A second and related reason is that participation in state policy processes has increased. Technically, federal statutes mandate notice and comment procedures wherein relevant parties are encouraged to attend public hearings. Environmental statutes mandating public participation include the Federal Water Pollution Control Act, the Resource Conservation and Recovery Act, and the Comprehensive Environmental Response.[11] Further, interest groups such as environmentalists are motivated to utilize these mechanisms at the state level. The responsibility of state governments for many issues of concern to environmentalists is widely recognized as "part of the legacy of Ronald Reagan's presidency."[12] Whether or not increased participation fosters responsive programs is an arguable prospect, but it does ensure that viewpoints other than those of traditional economic interests will be heard.

Third, following a brief period of fiscal restraint exemplified by Proposition 13 in California, some states responded to decentralization during the economic downturns in the early 1980s by levying more taxes and then, subsequent to economic recovery, found themselves with fiscal surpluses.[13] Those surpluses allowed state policymakers to increase expenditures at a faster rate than was occurring at either the national or local levels.[14] One can only take this argument so far, for the record of state governments in replacing the federal cuts of the Reagan administration was hardly uniform.[15] However, one could still argue that state policymakers did develop new sources of income, such as lotteries, which may enhance independence in future years.

Fourth, while mobility between state and national office has always been pursued by politicians, greater visibility and more extensive media attention in recent years has enhanced the importance of a state official's record. One need look no further than the candidacy of Massachusetts governor Dukakis in his 1988 presidential bid for evidence of this phenomenon. The increased visibility of subnational policymakers provides a rational reason to expect ambitious state officials to pursue responsible policies.

Fifth, the development of state political institutions has fostered competition in the policy arena. Competition often takes the form of innovation and new ideas, thereby providing policy leadership. Evidence for institutional growth is found in all branches. Governorships have centralized and consolidated. All but three states now have four-year terms for their governors and allow successive terms. Further, governors now have power to make more appointments and veto more legislation.[16] All but seven state legislatures now meet annually and have become more like the national legislature in terms of longevity, salaries, and committee structure.[17] State courts have become more active and more involved in numerous areas.[18] Finally, state bureaucracies have paralleled elected bodies in their growing sophistication and professionalization.[19] While the impact of each individual change is debatable, together they at least suggest increasing activity at the state level.

The state resurgence argument suggests reconsideration of the role of need in determining state behavior. If states have indeed developed responsive institutions, then at least some correlation between need and response could be expected. Several empirical analyses have dampened these expectations. Work on air pollution and hazardous waste policies in the late 1970s and early 1980s showed little evidence of "matching."[20] Analyses of hazardous waste and groundwater policies in the latter part of the 1980s also display a rather weak relationship between severity and state responsiveness.[21] Even early proponents of the state resurgence argument acknowledge that other variables besides policy needs, such as available resources, affect state behavior.[22]

This literature suggests wariness in heavy reliance on the matching hypothesis to explain state behavior. While some correlation between severity and response may be evident, state behavior is likely to be affected by other factors as well. Still, the need variable is important to this study for several reasons. First, some correlation between severity and response is expected, but that correlation may vary across policies depending on the influence of other factors that I discuss below. Variance in matching suggests the potential for influence from leading state programs on overall state behavior. Second and related, inclusion of the severity variable in the model should aid identification of the conditions under which states do respond to public needs. Third, leading state programs can be identified in those states that display responsive behavior to severe pollution problems. The need variable in this study is measured by actual levels of pollution and by the presence within the state of sources that contribute to pollution problems.

Alternative hypotheses of state behavior

As rational as the matching hypothesis may sound, various literatures suggest that other economic or political variables also affect state behavior. In fact, matching has rarely been a popular characterization of state behavior in American political analysis. Rather, state governments have been characterized as responding to factors other than need. Several of these alternative perspectives are considered below.

Paved with national intentions. If the road to hell is paved with good intentions, then the most critical view of state governments is that the road to federal policy failure is paved with state modifications of national intentions. This perspective suggests that private parochial interests at the state level modify national policy efforts to the point that policy outcomes reflect very little of stated intentions. Madison may have been the first to offer this viewpoint, but he was hardly the last.[23] Analysts since Madison have suggested that powerful private interest groups within the state determine state behavior in pollution control policies.

This interest group perspective is based on the expectation that state and local policymakers are indeed closer to the people, but also that certain groups of people utilize that proximity more effectively than do others. In particular, organized and powerful interests dominate policy at subnational levels. The discriminatory policies of the southern states as determined by powerful white factions exemplifies this indictment.[24] Subnational policymakers subvert federal intentions because, even if not part of the organized interests, they are part of the existing power arrangement that depends on support from powerful groups.[25] Some states are particularly susceptible to internal pressure and thus display relatively weak programs.

The important factor that varies between states in this perspective is the indigenous presence of relevant interest groups. Although interest groups were originally the focus of political scientists, recent economic models of subnational behavior display an awareness of the potential strength of this variable.[26] Interest groups can aid state policymakers by supplying information and resources as well as apply pressure by withholding, or threatening to leave with, the same. Thus the stringency of state regulatory policies varies inversely to the strength of indigenous economic interest groups. This result is quite the opposite from that postulated by the matching hypothesis.

In terms of pollution control policies, this perspective suggests that subnational policymakers avoid confrontation with polluters and instead design their policies (or avoid them altogether) to appeal to organized economic interests. The literature on hazardous waste cited above finds that the presence of waste-generating industries can have a strong influence on nonresponsive state behavior.[27] Since appropriate state leaders are unlikely to develop in areas of heavy pollution, according to this argument, implementation of national intentions is ineffectual, and innovation by state and local governments is nonexistent. The economic interest group variable is measured by the relative contribution of polluting groups, farmers or utilities for example, to each state's economy.

Money talks. The "affordability model" of federalism portrays state governments as heavily dependent on available resoures for programs.[28] Both economists and political scientists recognize the importance of resources available to state policymakers. States with more money can more likely afford essential services and can innovate with responsive new techniques.[29] This implies that states with wealthier citizens have a larger base from which to fund policy programs such as environmental protection.[30] States with greater resources may be less likely to depend on federal support for programs and may feel less compelled to compete for other sources of income.

This perspective suggests that leading states may develop in areas of heavy need, but only if those areas are states with abundant resources. One reassessment of the state resurgence argument by its own pioneers cites the importance of "resource richness" in determining state behavior.[31] Abundant state resources depend upon the relative affluence of the state's population. The most common measure for citizen affluence, and the one that this study uses, is per capita income. Hypothetically, higher per capita income leads to stronger state programs.

The impact of political culture. Some argue that each state's own political culture determines its policies. Political culture can be defined as "the particular pattern of orientation to political action in which each political system is embedded."[32] Presumably, states have a higher tendency to foster responsive programs if they enjoy dynamic institutions with high citizen awareness and professional avenues for concern. This argument suggests that states are more likely to have strong environmental programs if the citizens of the state are oriented to government involvement in important decisons.

This variable can be operationalized in several ways. One measure utilizes an index combining constituent awareness with party responsiveness so that an increase in either component makes the index higher. Constituent awareness is usually measured by voter turnout in state elections. Party responsiveness has been measured by the relative presence of one party in the state legislature or by competitiveness of the two parties.[33] The closer to an even split between the two parties in elected representatives, the more dynamic the culture.[34] A second way to measure political culture utilizes the argument that certain states are more conducive to public decision making because their citizens have a "moralistic" view of communal public interest. This view emphasizes public concern, welfare, and participation. States can be characterized as either moralistic, individualistic (market-oriented), or traditionalistic (status quo-oriented).[35] Finally, this variable can be measured in simple terms of partisanship based on the argument that Democrats lend more support to environmental efforts than do Republicans.[36] When Democrats dominate a state house or legislature, then hypothetically the state has more developed environmental programs. In the following equations, I rely mainly on the first operationalization. Where dramatically different impacts occur from other measures, they will be made explicit.

Federal aid. Many analyses of the federal system suggest that political influence on state programs can be supplied by the national government through the application or withholding of grants.[37] This variable is not unrelated to those above. For example, increased federal aid may make programs more affordable. Further, more dynamic political cultures may stimulate the flow of more federal funds into the state. Hypothetically, those states with the greatest federal support have the most developed programs. As a result, leading states are not determined by factors indigenous to the states but rather by the external influence of federal financial resources.

This potential impact is operationalized in this study with actual amounts of aid when available. Unfortunately, the data are often incomplete, especially since the Reagan administration reduced collection and reporting. Nevertheless, the variable is measured in actual dollars, usually per capita, contributed to each state for each specific policy.

Expected performance of the model

Understanding state leadership requires assessing the impacts of these hypothesized independent variables. Since each perspective contains some potential validity, all are incorporated into one model of state behavior, which can then be placed within the context of a federal system. Some of the most accurate current models of state behavior account for at least some factors similar to those described above.[38] Nearly all of the equations in the following chapters include measures of the severity of the problem, presence of relevant interest groups, relative affluence, political culture, and federal aid.

To summarize, I expect that no single variable in the model determines state behavior. As the literature on state variance suggests, states vary in their own capabilities and their susceptibility to pressures. For instance, one recent model categorizes states in terms of institutional capacity and dependence on federal aid in environmental issues.[39] I do expect that there is some correlation between need and response, particularly in leading state programs. However, that relationship over all fifty states could be weak and could be affected by the presence of the other variables.

I also expect that this model may not be enough to explain the total leadership role of state governments. The model describes variance between states within policies, but what can account for the differences in variance across policy lines? For example, do more states exhibit responsive matching in one policy than in another? Further, assessing the impact of state governments on national policy efforts requires consideration of not only the variance between states, but also the exceptionalism of some. If the efforts of leading states substantially exceed those of other states, then the potential impact of state leaders is significant. If state leaders can influence other state governments, then altering federal grants and modifying internal factors, such as resources, are not the only means to change the programs of weak states. State governments, even if their efforts are not uniform, can provide national leadership in policy responsiveness. Overall state behavior can be significantly affected by the actions of state leaders. The following section proposes a theoretical framework within which to place the model of state government behavior.

The Dimensions of Federalism

State leadership in national policies is affected by the horizontal dimension of interstate competition and the vertical dimension of federal involvement. This section describes these dimensions of federalism as inherent in the definition of subnational governments in a federal system.

States in a federal system are political entities that differ from a national unit in two important respects. First, they number more than one. If only one existed, then indeed that state would be the national unit. Because more than one exists, the potential is created for horizontal competition between units. Competition may be over valuable resources and/or the import/export of commodities, including externalities.[40] Second, states are subnational entities. When united in any fashion, coordination and cooperation is possible

Potential interstate competition

	High ⟷ Low	
High ↑	point source water	mobile air
Low ↓	stationary air	diffuse water

Potential vertical influence

Figure 1.   Dimensions of Federal Policies affecting State Leadership

through top-down direction. Thus the potential for vertical involvement and subnational reliance on federal initiative is created.

These two dimensions of federalism are depicted in figure 1. The axes should be thought of as scales rather than dichotomous variables. My thesis is that these dimensions significantly affect the leadership and coordination of state efforts. As discussed below, the effects of the dimensions are reciprocal. The horizontal dimension of interstate competition influences vertical relations between states and the national government. The vertical dimension of federal involvement affects horizontal cooperation across state lines. The primary purpose of the typology is to facilitate systematic analysis of state behavior under different circumstances. To preview the overall organization of the work, I have placed the four pollution control policies to be examined within the four cells of the figure. This placement is based on characteristics to be described shortly.

Horizontal competition

The dimension of horizontal competition is inherently important in the American political system. In most of this analysis, this dimension refers to competition over economic resources rather than competition over innovations. Recognition of potential and real competition over resources between states actually predates the Constitution. Shapers of that document recognized that the Articles of Confederation provided few mechanisms to restrict state competition for western land and foreign commerce.[41] The Constitution created a framework for influencing competitive state behavior but retained considerable state autonomy and thus the potential to compete.

Each policy area involves some potential for competition. As described below, the subject of the policy itself—polluters or pollution in the context of this book—may be more or less likely to cross state boundaries. Increased and meaningful mobility enhances state competition. If the federal government were to specify all possible behavior by the states, then state action would be uniform and would preclude competition. This is, of course, highly unlikely. As many political scientists have explained, strong incentives exist for national policymakers, especially members of Congress, to pass vague legislation and leave details to some other entity.[42] Regulatory statutes, in particular, "tend to be vaguely drawn."[43] While the literature has concentrated mostly on national bureaucracies filling in the details, the task often falls to state governments.

Within each policy area, states may simply comply with stated federal requirements and utilize discretionary behavior to compete with their neighbors for economic resources. Or state policymakers may exceed minimum federal expectations with more stringent regulations, more innovative procedures, and more experimental approaches. State leadership involves the willingness of state governments to supersede explicit federal standards. I argue that the potential for horizontal competition within a policy significantly determines the willingness of state policymakers to exceed federal guidelines.

This argument is based on recognition of the power of interest groups as they exercise options of exit and voice.[44] Exit is the ability of an individual or a group to leave a market or political district. Voice is the ability to demand compliant behavior by some authority. The application of these concepts to public policy in a federal system is based on simple concepts of supply and demand. States supply public goods. Constituents demand policies that provide their preferred mix of goods. According to the famous Tiebout Hypothesis, "The consumer-voter may be viewed as picking that community which best satisfies his preference pattern for public goods."[45] Complete mobility, if it could occur, would enable an optimal match of consumer preferences and public policies, an ideal outcome known as "fiscal equivalence."[46]

Incomplete mobility produces a more realistic, and perhaps less optimal, outcome. In many cases, constituents would rather not move and so attempt to use the threat of relocation to encourage states to adjust public policies closer to their preferences. The threat of exit enhances the power of voice for those interests. National policy intentions may be relegated by states to secondary status behind the immediate demands of important constituents. For example, states may compete for industries through lax pollution control policies.

State behavior in regulatory policies such as pollution control is particularly sensitive to horizontal competition because, by definition, these policies impose costs and benefits, thereby creating "losers" and "winners."[47] If regulatory policies are implemented at subnational levels, then the threat of relocation is a potentially significant tool of leverage for targeted "losers" of policies. In particular, most regulatory, especially pollution control, policies target certain elements of business and industry.

The argument that these groups use the threat of relocation to alter subnational behavior can take either of two forms. The strong form asserts that businesses do and will relocate to areas where policies match their preferences, thus creating incentives for state policymakers to alter behavior. The milder argument, one of anticipated reactions, suggests that whether or not businesses do relocate, state policymakers behave as if they might. In other words, the threat of exit makes the voice of demands even louder.

Evidence for the strong argument has been, for the most part, negative. Although the occasional study has suggested that consumers do "vote with their feet," most studies of business regulation find little evidence for industry relocation based on specific policies.[48] In particular, these studies admit that business location decisions are much more likely to be determined by factors such as labor and markets rather than by specific subnational policies.[49] Even translating stringent regulations into taxation policy (to fund regulatory efforts) does not, according to some studies, significantly increase the effects.[50] A Conservation Foundation project on industrial plant siting, for example, concluded specifically that environmental laws do not drive industry from one state to another in search of lax standards.[51] In addition to this empirical evidence, a theoretical argument suggests that, analogous to national legislators, state policymakers respond only to their immediate (i.e., district) constituents and are therefore not influenced by threats to the economy of the whole state.

On the other hand, several factors temper these negative results, a tempering that motivates consideration of a milder alternative. First, these studies are usually based on interviews of business, not political, decision makers.[52] The results, therefore, do not explicitly reflect the behavior of states. For example, the Conservation Foundation study concentrated mainly on industrial decision making and very little on state actions.[53] Second, the breadth of these studies obscures the possibility that for some firms, specific policies (such as OSHA regulations for a labor-intensive firm) may be crucial.[54] Third, the argument that state politicians will hold little regard for the relocation effects on the whole state is debatable. Much of state policy-making is determined by the executive branch, which by definition must be concerned with the entire state. In addition, ambitious state politicians must necessarily rely on the record of their entire state to support claims of effective prior service.[55]

The milder alternative suggests the possibility that state policymakers perceive the potential of business relocation and modify state policies accordingly. In other words, even if it is true that few industries base relocation decisions on differentiated policies, that does not mean that states do not compete through such policies. Most states desire at least some economic development, and nearly all states do have offices that are specifically responsible for it. Even if efforts to attract external businesses are unsuccessful, the practice continues.

Indeed, states can and do offer location incentives such as tax breaks, tax-free industrial development bonds, and tax-free financing for pollution control equipment.[56] According to one survey, competition through these incentives has intensified dramatically during the 1960s and 1970s.[57] Regardless of the success of these efforts, certain pressures exist to maintain the competition. First, the occasional story of a business being enticed by local policies enhances the perception that such behavior is widespread. Such perceptions are only reinforced by publications advising businesses as to relocation decisions that utilize stringency of standards as a significant criterion.[58] Second, all parties are hesitant to take the first step toward moderation of behavior. Each state thinks that if it reduces costly development incentives, it may lose important constituents in the time it takes for their neighbors to follow suit. Further, state policymakers fear placing their own industries at a competitive disadvantage to industries in neighboring states.[59] Finally, within state governments, bureaucracies such as the economic development agencies press for continuing such behavior simply to keep themselves in operation.

In addition to recognizing that states may compete over resources, a theory of state behavior that accounts for interstate exit recognizes that interaction between states may also include competition over territorial rights. Exit in this case can take the form of one state's produce. Indeed, state borders do not obstruct the flow of commodities such as pollutants that are carried by air and water. Thus the quality of the air and water in one state may be affected by the external produce from neighboring states. The transfer of unwanted contaminants across state lines may do as much to increase the tension between states as does competition for businesses and industries.

States may interact by designing their policies in response to the actions of their neighbors or precisely because their neighbors cannot respond. The latter occurs when externalities cross state borders. If a state can enjoy the benefits of some activity while exporting the negative consequences, then its own perceptions toward that activity may be affected. For example, a polluting state may maximize its own benefit-cost ratio by placing all of its factories just inside its downwind and downstream borders.

In this analysis, pollution control policies are differentiated according to the potential for polluter relocation and interstate transfer of externalities. Together, these two phenomena suggest that the greater the potential for horizontal competition between states, the more likely it is that states will simply do what they have to in response to federal demands. State leadership in supersedure of minimum federal demands is more likely when interstate competition is less prevalent.

Vertical involvement

The second dimension affecting state behavior inherent in a federal system is that of vertical involvement. Even the loosest confederation of subnational units creates some mechanism that may exert vertical influence on its members. Political scientists have long realized the importance of vertical influence. Madison urged extension of the vertical sphere to involve a greater number and diversity of interests and thereby avoid "tyranny of the majority."[60] Political analysts since have described and debated shifting trends of national involvement in state behavior.[61]

Each policy area involves a different type of vertical involvement. As described below, the legislation and statutory language of the policy mandates the amount and sort of federal influence on state behavior. At one extreme, when the national government is completely removed from a policy, states act completely autonomously. Such a possibility is no longer likely as the national government has become more involved in most domestic policies over the last fifty years. At the other extreme, the federal government is so involved in state decisions as to preempt state autonomy. This is also unlikely and becoming even more so. Not only would complete preemption be counter to two centuries of tradition, but national assumption of state expenses is a dubious prospect in a time of increasing federal deficits. Thus vertical involvement varies between these two extremes. Policies are differentiated in this book according to the degree of vertical involvement.

Vertical involvement affects state leadership. State leadership involves cooperation in, competition over, and dissemination of innovations and effective practices. Federal intervention can facilitate coordination and communication of state efforts. The process can involve signaling, cue taking, information exchange, development of professional associations, and coordinated research. Diffusion of progressive efforts is crucial to the evolution of progress in a federal system. In fact, the seminal piece in the literature on innovation suggests that "this process of competition and emulation, or cuetaking, is an important phenomenon which determines in large part the pace and direction of social and political change in the American states."[62] That work and other previous efforts analyzed variation in innovation across states. I emphasize variation across policies. I argue that the greater the extent of vertical involvement in a policy, the quicker the coordination of leading state efforts. Change through dispersion and emulation of innovation is fostered by open lines of communication that can be maintained by vertical hierarchy. Without vertical facilitation of change, rapid development of programs by all states is unlikely.

Why would autonomous state behavior not result in cooperation and coordination? The theoretical basis for this argument derives from the concept of power in a narrow setting.[63] The more narrow the scope of conflict, the more advantage to the organized interest and the more pressure to maintain the status quo. Pressure for the status quo from private good interest groups, such as business and industry, is theoretically more effective at the state level than pressure for change from public good groups, such as environmentalists. Political scientists have argued that subnational policymakers are particularly susceptible to the pressure from those interest groups involved in the policy to be administered.[64] Pressure can take the form of lobbying, campaign financing, providing information and resources, employment opportunities, and the breeding of familiarity. These pressures can be so strong, particularly at subnational levels, that policymakers may become "captured," serving rather than regulating private interests.[65] One empirical example of behavior to avoid potential capture at subnational levels is the mandating of regular transfers of agency personnel in order to negate the development of close allegiances between employees and local groups.[66]

What advantages do private good groups enjoy over public good groups at subnational levels? Perhaps the largest advantage to private good groups at the state level concerns "free riders." "Free riders" are noncontributing members of interest groups who enjoy the fruits of others' labors.[67] For example, a doctor who is not a member of the AMA but who benefits from their achievements is a free rider. Free riders can sap groups of their greatest strength, the strength in active members. The potential for damage from free riders differs between private and public good groups at the state level for the following reasons.

For industry, the interest group is made up of organizations each motivated by the pursuit of higher profits. Since this is also the motivation for the interest group, no disparity exists between internal and external goals. Further, because the costs involved in regulatory policy are concentrated, a narrowing of the sphere diminishes the attractiveness of a free rider strategy for organizations. Those who do not utilize voice to attempt to influence policy are destined to suffer the costs of the regulation: lower profits. Businesses can and do free ride, especially at the associational level, but their policy preferences and impact on local economies are generally fairly obvious.

With diffuse benefits, however, public good interest groups face a different set of internal dynamics. The perceived efficacy of the organization must replace profits as the motivating factor for members' activity. Individual members of noneconomic interest groups are more willing to contribute when they perceive higher personal efficacy in striving for the group-pursued goal.[68] Those perceptions can be manipulated by group leaders. In a federal system, one powerful means for leaders to achieve a perceived level of importance high enough to attract and maintain members is to concentrate on issues in the national limelight. With national publicity, the goal itself attains greater significance and the individual member feels that his or her effort is more crucial. This is one reason why public good interest group activity is more likely to be concentrated at the national rather than at subnational levels.

Another difference between private and public good groups that also accrues to the former's advantage at subnational levels is the simple fact that the former are made up of organizations whereas the latter must become them. Businesses need worry less about creating efficient new lobbying organizations. Further, their efforts can be expressed through single representatives. Public good groups, on the other hand, must mobilize grass-roots support to legitimize their points of view. They must become organizations rather than funnel the preferences of existing ones. In doing so, public good groups recognize that they enjoy greater economies of scale by building one large organization at the national level instead of different ones for each individual state.

Conceivably, federal involvement can moderate the inertia created by private good group advocacy for the status quo in otherwise autonomous state programs. Each policy entails some form of involvement by the federal government. If that involvement enhances communication and coordination across state lines, then the efforts of leading state programs can be more rapidly emulated and disseminated. Why would leading states even bother to utilize the federal government to stimulate the performance of lax states? Lead states may desire to diminish the gap between themselves and their less stringent counterparts to negate competitive relocation advantages offered by states with lax regulation. Without federal mechanisms for dispersion of innovations, such diminution remains difficult.

Policy differentiation in a federal system

These arguments necessitate consideration of the impacts of the dimensions of federalism on state leadership. Accounting for the differences in the impacts of federal dimensions requires differentiation between policies. Policy differentiation has received increasing attention in the federalism literature. Whereas much of the early work on federal systems concentrated on broad characterizations of federal aggrandizement or centralization versus decentralization questions, some of the most revealing recent work differentiates policies by issue area and type of federal involvement.[69] In this book I differentiate pollution control policies according to the potential for interstate competition and the autonomy awarded subnational governments. Reasons for the categorization of each policy area will be given.

The Model in the Context of Pollution Control Policies

Hypothesized impacts of the dimensions of federalism

Perhaps because the emphasis on state leadership is so recent, many important questions remain unanswered. How important is state leadership? Do state leaders develop in areas of greatest need? Do state leaders develop policy beyond minimum federal requirements? Are innovations and developments in leading state programs disseminated and coordinated across state lines? I address these questions by examining a model of state government behavior in policies that are differentiated along dimensions of horizontal competition and vertical influence. I offer three hypotheses concerning the leadership role of state governments. The first measures the prevalence of state leadership. How many states show responsive behavior in terms of overall matching between need and response? The second and third hypotheses address the actions and effects of leading state programs. Do they innovate? Do they coordinate their efforts?

Hypothesis 1. Some correlation between severity and response will be apparent at the state level, but overall matching will vary according to the dimensions of federalism:
a. when vertical involvement is low and horizontal competition is high, matching will be less apparent;
b. when vertical involvement is high and horizontal competition is low, matching will be more apparent;
c. when vertical involvement and horizontal competition are both high or low, matching will depend on other variables.

This hypothesis tests the consistency of responsive state behavior. What is the extent of state leadership? The hypothesis acknowledges the finding in the literature that states vary in terms of their responsiveness to problems, but argues that variance differs across policy lines. In the extreme, if vertical involvement were absent and horizontal competition were dominant, the variance between states would be greatly exaggerated by the presence of powerful economic sectors. If those sectors create policy needs, as in air and water pollution, matching would be nonexistent. On the other extreme, if vertical involvement were complete and horizontal competition nonexistent, then the variance between states would be largely determined by the federal government. If national intentions emphasized policy responses, as in air and water pollution, then matching would be quite high. Between these two extremes, the relative impact of federal dimensions affects the prevalence of state responsiveness. When the dimensions balance, some matching is likely, but overall state behavior is influenced by other variables as well.

Hypothesis 2. The lower the level of interstate competition in a policy area, the more likely that leading state programs will supersede federal guidelines.

This hypothesis concerns the exceptionalism of leading state programs. What is the depth of state leadership? The horizontal dimension of competition hypothetically influences a vertical state response, the willingness to exceed federal standards. When policies involve less competition for states from their neighbors, leading state policymakers may be less hesitant to surpass minimum expectations. Supersedure of federal guidelines includes innovations, experimentation, and the utilization of greater stringency than required.

Hypothesis 3. The higher the degree of vertical involvement in a policy area, the greater the dissemination and coordination of leading state efforts.

This hypothesis addresses the influence of leading state programs. What is the impact of state leadership? The vertical dimension of federal influence hypothetically affects a horizontal phenomenon, diffusion of leading efforts across state lines. When policies enable states to utilize the federal hierarchy, leading state policymakers can more easily share their efforts. Diffusion includes interstate agreements, seminars, imitation, and emulation.

The context of pollution control policies

Testing of these hypotheses is conducted within four specific pollution control policies. The four policies are shown within the cells of figure 1. These policies are selected for two reasons. The first reason is substantive. Air pollution is emitted from both stationary (factories, utilities) and mobile (automobiles, trucks) sources. Water pollution is discharged from both point (factories and municipalities) and nonpoint (diffuse such as farms and mines) sources. Each category constitutes roughly half of the pollution load for that particular medium. For example, diffuse sources of water pollution contribute roughly half of the contaminants now fouling American waters. Each of these four categories is addressed by a different set of national and state programs. Each category involves serious pollution problems.

The second reason is theoretical. Environmental policy has been and will continue to be an important issue area for state behavior. It thus can serve as a signpost for judging state performance.[70] These four policies cover a range of circumstances that vary along the dimensions of federalism that have been described. Those dimensions are depicted as the axes of the two-by-two cell in figure 1. Placement on the horizontal axis is determined by the nature of the pollution and categorizes policies as involving high or low potential for competition over polluters and pollutants. The vertical axis categorizes vertical involvement in specific pollution control policies as high or low. Placement is determined by the degree and type of federal intervention in state policies.

Pollution control policies are appropriate to this study of state leadership because they are comparable in so many ways and yet differ along dimensions that are important to a study of federalism. The next section briefly describes the existence of pollution control policies and state behavior prior to the differentiation and emphasis given them in the late 1960s and early 1970s.

Pollution control policies prior to 1970

Prior to the environmental movement of the late 1960s, the pollution control policies that did exist in the United States were the nearly exclusive domain of state and local governments. With the federal role limited to funding and advice, environmental policies varied widely between states in terms of stringency and effectiveness. Overall, states made little systematic progress in the fight against pollution. In particular, many states feared the threat of relocation of important constituents and thus avoided stringent regulations.[71]

State and local officials, enticed by federal dollars, gradually became more receptive to a federal presence in these policies. The increase in federal financial involvement in the air pollution and sewage treatment policies of the states is reflected in tables 1.1 and 1.2. Efforts to expand the federal role beyond financial assistance to the level of standard setting before 1965 in both air and water issues were unsuccessful due to fierce opposition by states and industries and claims of violations of states' rights by members of Congress.[72]

The year 1965 marked a turning point in the relationship between the national and state governments in environmental issues. Although federal jurisdiction was still limited to interstate waters, the Water Quality Act of 1965

Table 1.1   National Air Pollution Control Resources

Year (FY)	Authorization ($ millions)	Appropriation ($ millions)	End of Year Employment
1955	—	0.2	29
1956	5.0	1.7	94
1957	5.0	2.7	157
1958	5.0	4.0	213
1959	5.0	3.7	218
1960	6.0	4.2	251
1961	5.0	5.9	296
1962	5.0	8.8	371
1963	5.0	11.1	407
1964	10.0	13.0	414
1965	25.0	21.0	525
1966	30.5	26.7	697
1967	46.0	40.1	888
1968	109.0	64.2	1070
1969	185.0	88.7	1065
1970	179.0	108.8	1016
1971	220.0	107.8	1261

Source: Originally unpublished table (January 1971) from the National Air Pollution Control Administration as printed in Jones, *Clean Air*, p. 113.

stipulated the withholding of federal funds if states did not establish standards for those bodies of water by a certain deadline (mid-1967). Standards were also the focus of the 1965 Motor Vehicle Air Pollution Control Act. Following the lead set by the state of California, the national government determined to establish emission standards for new automobiles. Significantly, the automobile industry did not provide the resistance that in the past could have obstructed the implementation of national standards, perhaps largely because "it feared fifty diverse state standards far more than a uniform federal standard."[73]

Federal involvement in these policy areas continued to increase in the next few years. The 1966 Clean Water Restoration Act included a large allocation of federal funds (table 1.2) as well as placement of the Federal Water Pollution Control Administration in the Interior Department instead of HEW, the former being less dependent on state initiative than was the latter. The 1967 Air Quality Control Act also greatly increased federal financial involvement (table 1.1) in air policy.

The increased federal role resulted from widespread and growing perceptions that the states were simply not doing the job. Between 1965 and 1970

Table 1.2   Sewage Treatment Facilities

Year (FY)	Authorization ($ millions)	Appropriation ($ millions)	Actual Spent ($ millions)
1957	50	50	38
1958	50	46	47
1959	50	47	—
1960	50	46	—
1961	50	46	45
1962	80	80	65
1963	90	90	92
1964	100	90	85
1965	100	90	85
1966	130	121	—
1967	150	150	—
1968	450	203	—
1969	700	214	—
Total	2050	1272	—

Source: Rohrer, *The Environment Crisis*, pp. 131–32.

the percentage of people who ranked the reduction of air and water pollution as one of the three national problems most in need of redress jumped, according to Gallup polls, from 17 to 53.[74] A more specific breakdown, from Opinion Research Corporation, is presented in table 1.3.[75] The table shows growing awareness of the severity of pollution problems throughout the country.

Increasing concern for the environment was matched by accelerated demands for governmental response. Both phenomena climaxed on 22 April 1970 with Earth Day. Conceived by Senator Gaylord Nelson (D-Wisconsin) as a teach-in similar to those being held on Vietnam, the event was remarkably successful in terms of media coverage and the numbers of people involved.[76] Nelson himself confided in an interview that the purpose of the event was to secure a position for environmental issues on the national political agenda. He recounted the disappointment at the time in state handling of environmental issues. Without federal involvement, he recalled thinking, "you'd have a hell of a mess."[77]

Differentiation between specific pollution control policies

Throughout the two decades that followed the first Earth Day, specific pollution control policies have been formulated at different times and with different emphases. I categorize those policies according to the dimensions of

Table 1.3   Concern over Pollution

Compared to other parts of the country, how serious, in your opinion, do you think the problem of air/water pollution is in this area—very serious, somewhat serious, serious, or not very serious?

	Air	Water
By time	*Very, somewhat serious*	
May 1965:	28%	35%
November 1966:	48	49
November 1967:	53	52
November 1968:	55	58
June 1970:	69	74
By geographic region	*Very serious*	
Northeast:		
1965	20	21
1967	29	30
1968	34	35
1970	51	53
Midwest:		
1965	8	14
1967	29	28
1968	26	35
1970	33	41
South:		
1965	3	6
1967	14	18
1968	12	18
1970	20	27
West:		
1965	13	6
1967	42	17
1968	37	22
1970	42	28

Source: Opinion Research Corporation surveys reprinted in Hazel Erskine, "The polls: Pollution and its costs" *The Public Opinion Quarterly* 36, no. 1 (1972): 121–22.

federalism. I describe that differentiation briefly here before asserting the need for such a component in the current literature.

Potential horizontal interactions between states are determined by the subject of attention of the individual policy. Policies affecting point sources of water pollution and stationary sources of air pollution involve high potential for externality transfers and relocation threats. Policies that affect mobile sources of air pollution and nonpoint sources of water pollution entail lower levels of interstate competition for two reasons. First, the pollutants themselves are not released at high levels of the atmosphere nor directly into strong currents of water. Second, these pollutants are released by more numerous, more diverse, and less mobile (as collective units) sources of pollution.

The degree and type of vertical influence is created by the policy designs specific to individual aspects of pollution control policies. State behavior regarding point source water and mobile source air polluters is subject to severe federal influence in means of attainment as well as in goals. Regulation of stationary source air polluters and nonpoint source water polluters involves more autonomy available to state policymakers as the federal role rarely exceeds that of goal setting.

The literature on environmental issues in the United States has offered considerable study of state governments but rarely comparative analysis of state behavior that accounts for differentiation across policy lines. Recent works have concentrated on the reaction of state governments to the Reagan administration's decentralization efforts,[78] the overall "resurgence" of state governments,[79] or state performance within individual policy areas.[80] Some sources have provided data on the behavior of all fifty states on different issues.[81] However, those data are presented with little theoretical differentiation of policies. As Lester and Lombard argue, the literature would benefit from more compelling theoretical arguments, greater awareness of intergovernmental relations, less emphasis on single points in time, and concurrent utilization of both fifty-state data analysis and in-depth case studies.[82] As the following section explains, I respond to these suggestions with an analysis that differentiates between policies on theoretical grounds of interstate and intergovernmental dimensions, that considers progression of state behavior over time, and that uses fifty-state data as well as specific case studies of individual states.

Design of the Study

The next four chapters discuss the effects of the dimensions of federalism on state leadership in each of the four policy areas. Included in these analyses are descriptions of policy design differences, explanation for those differences, identification of effective state programs, and consideration of evidence for the hypotheses developed in this chapter. Within each area of study, I describe a specific state program that exemplifies the lead role that the states can take in these issues. The concluding chapter summarizes and compares the results of the policy case studies.

I measure state behavior with both qualitative and quantitative data. I have conducted numerous interviews with state policymakers. My interviews focused particularly on bureaucrats and administrators because they are actually implementing the programs at hand. Other studies have shown that state environmental officials are usually the most influential actors in the implementation of pollution control policies.[83] This work also utilizes a variety of sources of data on state environmental programs, but concentrates on those that are specific to individual policies.[84] These measures include data on financial expenditures, compliance, attainment, procedures, indices created from numerous sources specific to each program, and more summary measures such as rankings of state programs provided in reports by the Fund for Renewable Energy and the Environment (FREE).[85] The Fund is a nonprofit, educational organization that collects and compiles data on state environmental behavior.[86] For the most part, the fifty-state, quantitative data are used to address the matching hypothesis. The interview data focus more on the behavior of leading state programs.

The states selected for in-depth analysis were chosen because they were state leaders in specific policies. Details for those choices are described in each chapter. The four states represent wide geographic variance being from the Midwest (Wisconsin), the South (North Carolina), the West (California), and the Plains (Iowa). Further, the four states studied vary widely on the dimensions in recent models of state institutional capacity and dependence on federal aid.[87] The states also represent a range of reputations as far as previous innovative behavior. According to Walker's innovation index, the four states mentioned above ranked, respectively, third, tenth, twenty-fourth, and twenty-ninth. Only the first two were portrayed as regional pacesetters.[88] Finally, the states have varying political histories and demographics, as described in the following chapters.

The preceding identification of state leaders motivates one disclaimer for the work as a whole. The common thread between the four state programs studied in depth is that they are exceptional. Positive descriptions of aspects of those programs should not be taken to imply that all state governments have pollution control well in hand. Rather, consideration of the experiences and efforts of state leaders may be helpful in understanding the effects of state leadership on public policy. My intent is not to stimulate excessive optimism about pollution control in the United States. Rather, I have two general purposes.

First, numerous observers, ranging from academics to politicians to journalists, have criticized the failure to achieve the lofty goals set for environmental policies in the early 1970s.[89] If indeed the behavior of the states is crucial to policy attainment, then closer attention needs to be paid to state assignments and discretions in any attempted remedies to pollution control policies. Support from this analysis for the specific hypotheses tested might motivate further examination of the role of state leadership in the implementation of these and other federal policies. Two early pioneers of the work on the revitalization of state governments closed their book with the comment that "Our belief is that the states will take the lead in these [policymaking] endeavors."[90] This work considers the meaning of that leadership.

The second purpose is related but concerns a broader potential conclusion. This study shows that in defining policy outcomes, it is not enough to say that institutions matter, or that states matter; rather, we must acknowledge that the institution of federalism itself matters. This only becomes more important as greater responsibility, in all domestic policy areas, is delegated to the state level. This book offers horizontal and vertical dimensions of federalism as a theoretical framework for further analysis of public policy in a federal system.

2 Stationary Source Air Pollution Control

In 1983 James Rickun was appointed head of the Wisconsin Air Toxics Task Force. The ATTF consisted of environmentalists, industrialists, and public officials. Their mission was to formulate a state rule for the control of toxic air pollutants not covered by federal legislation. By the time the mission was completed several years later, Wisconsin had one of the first and one of the strongest toxic rules in the country. Rickun himself moved on to another job, exhausted by both the intensity of debate within the ATTF and pressure from elements of the community.

The story is illustrative of several aspects of stationary source (SS) air pollution control in the United States. First, much initiative is left to state governments and some states have provided the lead in pollution control. Second, the variance (as shown by the scarcity of toxic rules) between states in this policy area is quite high. Third, the stakes are large and the pressure on state officials is intense. Thus states may promulgate some regulations in a federal vacuum, but they are wary of exceeding existing federal standards. Further, coordination across state lines is rare to the point of being problematic. This chapter examines state behavior in SS air pollution control. It considers the role of states in this policy area, the overall behavior of state governments, and then returns to the case study of Wisconsin to discuss the efforts of a leading state program.

The Role of the States in Stationary Source Air Pollution Control

The role of the states in this policy area is largely established by the Clean Air Act (CAA) of 1970. This legislation constitutes a dramatic shift in federal-state relations in American domestic policy. The law calls for national establishment of state goals and calls federal financial assistance and leadership "essential."[1] Nevertheless, the importance retained by the states in the Act is evident in the statement that "the prevention and control of air pollution at its source is the primary responsibility of States and local governments."[2] Thus while the federal government is responsible for goals, the state governments are responsible for means of attainment of those goals. The history of state behavior in the control of ss air pollution is one shaped by this relationship.

Stationary source polluters are buildings, structures, facilities, or installations that have the potential to emit pollutants subject to regulation by the CAA and that are fixed to specific sites. The CAA differentiates between mobile source and stationary source control. Whereas automobile emissions are susceptible to uniform national standards (chapter 4), ss air pollution is much more locally specific. Thus mobile source pollution control involves a relatively direct relationship between the national government and mobile polluters, whereas ss control is implemented through state and local governments. Other reasons for this differentiation will be discussed shortly.

The precedent for the ss arrangement in the CAA was established by 1967 air legislation whereby states would set their own standards for regions by sources (or classes of sources) of pollution, subject to HEW approval, and devise enforcement to meet them. These regions would be established on topographical and meteorological as well as political bases. For instance, West Virginia had ten regions.[3] This 1967 legislation thus necessitated some federal involvement but "rested heavily on state initiative."[4]

The 1967 legislation was vague and unenforceable. Neither federal nor state progress toward controlling air pollution achieved the goals of the legislation by 1970. Not even half of the planned ninety-two air control regions were designated by July.[5] Only twenty states adopted comprehensive pollution legislation and just six more strengthened previous pollution control efforts.[6] Data in the Presidential Message on the Environment in early 1970 indicated how unlikely it was that objectives would be met. Only six of the fifty states were spending more than the annual per capita expenditure (twenty-five cents) estimated to be needed to control air pollution. Also, less than 50 percent of state agencies had as many as ten positions filled in air pollution control capacities.[7] Such slow progress stimulated demands for change in 1970.

Low vertical involvement

The 1970 Clean Air Act Amendments represent change in many ways, but they do retain the states' role in implementation and enforcement. More federal authority was instituted with the required setting of national primary ambient air quality standards by the administrator of the new Environmental Protection Agency (EPA). However, each state would adopt its own plan to meet and enforce these standards. SS air policy is thus based on a two-tiered structure of federalism.

First, federal goal setting is required through EPA standards for the following criteria pollutants. Carbon monoxide is a colorless gas resulting from incomplete combustion, mainly in automobile engines, that can enter the bloodstream and cause dizziness, headaches, and impaired heart functions. Hydrocarbons are compounds of carbon and hydrogen that are released through the burning of fossil fuels. Nitric and sulfur oxides are substances that can combine with oxygen to weaken pulmonary functions and lower resistance to disease. Particulates are tiny particles of contaminants that can enter the lungs and cause respiratory problems. In addition to these specified pollutants, the EPA is authorized to regulate unforeseen hazardous pollutants. This requirement involves setting primary standards for the health of the most sensitive groups in the population and secondary standards for the public welfare. The primary standards were to be met throughout each air quality control region by 1982 or 1987, depending on the pollutant.

The federal government is also required to establish New Source Performance Standards (NSPS) for new polluting industries whereby "All new sources must meet engineering-based standards set by the EPA for each source category in each industry . . . on a uniform national basis."[8] The NSPS provision constitutes serious encroachment on state prerogatives, but it is important to note that NSPS sources constitute roughly one-tenth of those major sources monitored by the EPA.[9]

The second tier of the structure of SS air pollution policy reflects the retention of traditional notions of federalism. Each state is required to develop a State Implementation Plan (SIP) to provide for the implementation and enforcement of standards within each of the air quality control regions within the state (247 total nationwide). The use of air quality standards instead of emission limits "gives the states the opportunity to choose any pollution control mix adequate to meet these standards."[10] This mandated flexibility should not be underestimated. The possible methods available for sampling and analysis are complex and varied.[11] The EPA can disapprove an SIP if deemed inadequate and can even offer its own plan for that state. However, it cannot compel state enforcement of an SIP but can only utilize indirect means such as the withholding of federal grants. Although not required to, most states have based their standards on particular technologies and still utilize renewable operating permits for old as well as the required new sources.[12]

This two-tiered structure constitutes a radical, but not complete, change in patterns of cooperative federalism in environmental policy. At the time of passage, federal specification of state objectives was recognized as a serious break from traditional assignment of state responsibilities that had been honored as late as the 1967 air legislation. In 1970 testimony, Senator Cooper of the Senate Air and Water Pollution Subcommittee described the CAA as moving "a long way toward national emission standards—a concept rejected by the committee in 1967."[13] Federal intrusion on state prerogatives was justified by the emphasis on public health, a traditional federal concern, operationalized in the ambient goals. In recent interviews, however, Cooper and chief staffer Leon Billings, who wrote much of the legislation, both recalled awareness of federalism as an institution that should only be altered incrementally. Thus the legislation stopped short of specifying means as well as goals of attainment for the states.

The high degree of discretion awarded state governments in this policy area was strengthened somewhat in 1977 amendments to the CAA. Other than the specific aspects related to interstate interaction described later, these amendments "confer much greater authority on state and local governments, particularly in such areas as the issuance of variances and the extension of compliance schedules."[14] One exception is that the 1977 amendments explicitly allowed imposition of conditions on certain federal grants to ensure state compliance. The amendments authorized cutoff of not just air policy grants, but also transportation (highway) grants to recalcitrant states. This federal threat has remained just that as extensive litigation has tied up the issue in courts and rendered the conditionality virtually ineffective. For example, even as cities drifted toward nonattainment of deadlines, by 1987 only Cleveland had suffered any EPA-imposed cutoff of highway money as a sanction.[15]

High state autonomy in SS air policy was not reduced by the Reagan administration. Reagan pushed for serious changes in clean air legislation and attempted, somewhat ironically, to use federal sanctions to achieve revisions that would grant even greater discretion to state authorities. EPA Administrator Anne Burford threatened to impose mandated grant cutoffs on noncompliant counties in 1982, thereby forcing Congress to rewrite the law to reduce federal authority. These threats inspired members of Congress to impose a moratorium on EPA sanctions, which contributed to the stalemate over clean air legislation in subsequent years. Nevertheless, the Reagan administration has been able to increase state authority while reducing state funds.[16] For example, regulation of air toxics has been largely delegated to state governments.[17] Rather than assuming an extensive role in toxic regulation, the EPA relies on states to develop their own toxic air pollution pro-

Table 2.1   Air Pollution Control Funding (million 1972 dollars)

	Federal				State and Local				
	Abate & Contrl	Reg & Monitr	Res & Devel	Total	Abate & Contrl	Reg & Monitr	Res & Devel	Total	Gov Ent Capital
1972	56	48	104	208	1	95	17	113	63
1973	45	47	118	210	1	107	6	114	82
1974	44	45	86	175	1	112	6	119	102
1975	61	53	84	198	1	109	6	116	100
1976	68	52	95	215	1	98	5	104	150
1977	61	57	98	216	1	109	5	115	188
1978	48	62	92	202	1	116	5	122	193
1979	48	63	65	176	1	118	5	124	265
1980	39	71	72	182	1	109	3	113	378
1981	31	57	66	154	1	108	1	110	406
1982	31	46	58	135	1	101	1	103	441
1983	45	41	61	147	1	99	2	102	369

Source: U.S. CEQ, *Environmental Quality* (1983), pp. 616–18.

grams. Simultaneous with the increase in state authority for programs like toxics, state air grants, according to the Congressional Budget Office, were cut by 33 percent between 1981 and 1984.[18] State autonomy was, if anything, therefore increased as a result of greater responsibility and diminished federal resources.

In 1989 President Bush proposed several changes in air pollution provisions that involve the states. The Bush plan singles out midwestern utilities for serious sulfur dioxide emission reductions (cut twenty million tons in half by the year 2000) and requires nonattainment areas to utilize vehicles that run on alternative fuels by 2004.[19] The Bush plan displays significant revisions from the proposals of his predecessor. Specifically, the plan emphasizes prevention as well as cleanup and encourages strict enforcement to attain compliance.[20] The revised CAA may also benefit from changes in Congress such as a new majority leader, which may help break the gridlock that has stymied renewal for years.[21] Nevertheless, state discretion in attainment remains guaranteed and federal commitment to these goals is yet to be demonstrated.[22]

Federal intervention in a policy area takes the form of resources as well as policy design. Table 2.1 details the amounts of the early federal contribution to air pollution control efforts. One might notice that although the decrease in federal funds was exaggerated by the Reagan administration, the decline actually began before then. This indicates the influence of economic circumstances on national behavior in federal relations. Stagflation made cuts essential even for the environmentally motivated Carter administration. Total state and local spending on air pollution remained fairly constant throughout the period.

Total amounts do not tell the whole story, obviously. In contrast to project targeting in water (which will be discussed later), federal financial aid in ss air policy was bequeathed to subnational agencies in order to assist their own relatively discretionary activities.[23] The EPA could grant up to two-thirds of planning costs and one-half of maintenance costs, fractions which could be increased for special programs to three-quarters and three-fifths. These grants enhanced the flexibility of air policy agencies at the subnational level. In addition, a General Accounting Office (GAO) audit in 1984 found that the EPA did not have sufficient procedures for ensuring that grantees maintained uniform levels of effort.[24] This is just one more indication of loose federal influence.

The allowance of state autonomy and discretion in implementation and enforcement since 1970 determines the characterization of vertical involvement as low in ss air policy. Federal control would be high if two factors were different. First, goals specified for the states would be so specific and precise that variances and externalities would not be possible. Second, goals would be meaningful in that noncompliance would stimulate automatic federal sanctions. Neither factor is evident in ss air policy.

High horizontal competition

ss air policy involves high potential for interstate competition both because of the subject of the policy and because of the policy itself. Both industries and their air pollutants can travel across geopolitical boundaries such as state lines. Recognition of potential interaction has been responsible for several changes in air pollution policy since 1970. This section describes those changes. They can be roughly categorized as affecting either attainment (air quality regions that have achieved goals) or nonattainment areas.

The most important change regarding attainment areas is the concept of Prevention of Significant Deterioration (PSD). The PSD policy was stimulated by awareness of potentially severe interstate competition. Inspired by the fear of degradation of "clean" areas down to the level of mandated standards, PSD resulted from a suit originally filed by the Sierra Club and twenty states (*Sierra Club v. Ruckelshaus*) in 1972. The suit was prompted by 1971 EPA guidelines that reversed previous statements and suggested that states should allow deterioration of air quality in areas that exceeded ambient standards.[25] Environmentalists feared that allowance of even marginal deterioration might foster intense state competition for polluting industries. As one state attorney general argued, PSD "prevents these areas from attempting to attract industry by enacting lenient emissions control requirements."[26] The original court decision in favor of the Sierra Club was upheld by the Supreme Court in 1973. The Court did not, however, define PSD or offer specifics. Subsequently, the EPA stepped in and issued regulations to implement the policy in August 1974.

When members of Congress legislated PSD, their actions also reflected concern for interstate effects. By the time legislation came up in 1976, representatives from states fearful of losing industry were more likely to support PSD than were those from "clean" states. In fact, the 1976 legislation was killed by a filibuster led by "clean" state Senators Garn (R-Utah) and Moss (D-Utah). Opponents of PSD openly admitted that their opposition was based on the fact that mandated nondegradation would deter the abilities of their states to develop by attracting outside industries.[27] The House version of the legislation was more lenient than the Senate one that inspired the filibuster. Fewer federal lands would be set aside as untouchable and states would be given more control in designating three classes of developable lands.

The PSD policy passed in 1977 closely resembled the 1976 House version. Three classes of attainment areas were established with specified allowable increments of pollutant increases in each. In addition, the conference compromise included a House allowance of SO_2 emissions to be exceeded in Class I areas for eighteen days a year if certain requirements (such as public hearings) were met.[28] Classification of lands into the different categories was left partially up to the states, but reclassification required federal approval. The ultimate resolution to difficult questions was and remains extremely complex, so much so as to make implementation difficult and criticism quite abundant.[29] Nevertheless, even this federal intervention into ss air policy retains considerable discretion for state governments.

Nonattainment areas also received increased attention during this time period. Again, political demands were based on the realization that ss air pollution involved potentially serious competition between states. Environmentalists were well aware that states could compete with each other for ss industries through lax pollution policies. Thus many advocated the allowance of new polluters in "dirty" areas only if their emissions were offset by reductions in existing pollution. Environmentalists also knew that nonattainment could be caused by pollution generated externally. They therefore promoted the use of scrubbers on smokestacks to reduce emissions.

The 1977 amendments adopted several measures for nonattainment areas. Congress mandated the installation of scrubbers, adopted provisions for emission offsets, and stipulated Lowest Achievable Emission Rates (LAER) for new polluters. The "bubble concept" involving emission tradeoffs within the same plant was endorsed and soon encouraged by the EPA for SIPs. Nonattainment deadlines were stretched to 1982 for states (five years after the original) and 1987 for cities. According to the 1977 Clean Air Act, states could choose either of two innovative procedures for nonattainment areas in future SIPs. The first allowed states to adopt the EPA offset policy whereby new polluting sources must obtain offsets from existing sources before beginning production. The second enabled states to facilitate growth by requiring existing sources to achieve more than just attainment levels so that a margin would be created for new polluters to enter the market.[30] Congress also recognized the problem of interstate pollution, particularly in "acid rain," and enabled the EPA to require pollution abatement in one state if a neighboring state petitions that its own air quality is being affected.[31]

These changes regarding nonattainment areas appear, at first glance, to be serious federal restrictions on horizontal interactions between states. However, the literary changes to Sections 110, 115, 123, and 126 were dependent on EPA enforcement that has been so inconsistent that the "substantive results . . . have been marginal, if not non-existent."[32] Further, models of the pollution process have simply been inadequate for states to prove that their problems were caused by specific external sources, proof necessary for succesful grievances.[33] Finally, even if models were more sophisticated, the EPA has resisted imposing serious restrictions on externalizing procedures such as tall smokestacks due to "heavy lobbying pressure from utility and copper smelting industries."[34]

The 1977 amendments represent an ironic manifestation of interstate competition in federal legislation that took on regional dimensions. The PSD provision, along with mandatory installation of scrubbers on coal burners, made industries more reluctant to relocate in "cleaner" areas of the country such as the Southwest.[35] These legislative mandates are an interesting result of the influence of an odd coalition of satisficing environmentalists and rustbelt industrialists. Nevertheless, loopholes and lax enforcement have allowed the interaction between states that was to be curbed to remain high.

The presence of high interstate competition in SS policy is the largest obstacle to overcome in passage of the recent Bush proposals. Control of acid rain, which will be discussed more fully later, has inspired such divisive regional competition in Congress that CAA reauthorizations rarely have made it out of committee in recent years. The Bush-proposed mandatory reductions of sulfur dioxide emissions from midwestern utilities and the allowed usage of low-sulfur coal rather than scrubbers suggest that West Virginia and some midwestern states will be the losers in competition over SS policy. This remains to be seen, however, as legislators from these areas could form a formidable coalition in the current Congress.

The Theoretical Model in the Current Context

Relative progress

By the 1980s, most observers agreed that progress in air pollution abatement had occurred, but not as fast as had been desired. According to the EPA, 93 percent of industries were in compliance with the provisions of the act by 1979.[36] A more detailed and recent summary is presented in table 2.2. Class A sources are those whose potential uncontrolled emissions are greater than 100 tons/year of any pollutant. NSPS refers to new polluting facilities. NESHAP accounts for hazardous air pollutants. The compliance rates are high. Many of these results are due to state behavior. In a 1985 report, the GAO examined six state agencies and concluded that they "were generally doing a good job in carrying out their air pollution control programs."[37] In 1986 the states conducted 28,463 inspections and source tests while the EPA ran 2,353 of their own.[38] These results reflect significant compliance with federal standards.

The effects on air quality have been substantial. Between 1960 and 1979, for example, particulates had decreased by 32 percent and SO_2 by 67 percent.[39] Some reduction in SS emissions since passage of the CAA is apparent in table 2.3. Other assessments have also cited some success in state air policy programs.[40] One analysis has shown that the stringency of state air programs is significantly correlated to the amount that emissions have been reduced.[41]

On the other hand, progress toward cleaner air has been painstakingly slow. Still more than half of the American people live in areas where standards for at least one of the targeted pollutants are occasionally violated.[42] The summer of 1988 was particulary rough when ozone levels rose an average of 14 percent. Even if standards are met, the health of many sensitive

Table 2.2   Stationary Source Compliance Record, 1986 (in %)

Source (n)	In Compliance	In Violation		Unknown
		On Schedule	No Schedule	
Class A (28,000)	92.5	2.1	4.1	1.2
NSPS (3,000)	91.0	1.9	4.9	2.2
NESHAP (1,000)	86.9	3.2	6.2	3.2

Source: U.S. EPA, *Progress in the Prevention and Control of Air Pollution in 1986*, p. VIII-1.

Table 2.3 National Air Pollution Emissions (million metric tons/year)

	Particulates	Sulfur Oxides	Nitrogen Oxides	Hydrocarbons	Carbon monoxide
1974					
Fuel Combustion, Stationary	6.2	24.1	13.1	1.6	1.4
Industrial Processes	10.4	5.9	.8	10.3	9.1
Transportation	1.3	.8	10.6	13.6	82.3
Other	1.3	.1	.3	5.6	8.0
Total	19.2	30.9	24.8	31.1	100.8
1977					
Fuel Combustion, Stationary	4.8	22.4	13.0	1.5	1.2
Industrial Processes	5.4	4.2	.7	10.1	8.3
Transportation	1.1	.8	9.2	11.5	85.7
Other	1.1	0	.2	5.2	7.5
Total	12.4	27.4	23.1	28.3	102.7
1980					
Fuel Combustion, Stationary	1.4	19.0	10.6	0.2	2.1
Industrial Processes	3.7	3.8	.7	10.8	5.8
Transportation	1.4	.9	9.1	7.8	69.1
Other	1.3	0	.3	3.0	8.4
Total	7.8	23.7	20.7	21.8	85.4

Sources:
1974—U.S. EPA, *1974 National Emissions Report*, p. vii
1977—U.S. EPA, *1977 National Emissions Trends Report*, p. vii
1980—U.S. EPA, *National Air Pollutant Emission Estimates, 1940–1980*, table 2-6.

members of the public is not protected. Many standards have yet to be established, including those for indoor pollution and dangerous toxics.[43] Last but certainly not least, acid rain continues to devastate statues, forests, and bodies of water in Canada as well as in many of the states. How many of these failures to meet expectations in ss air pollution control can be attributed to the states? Further, what can be expected from state leadership in this policy area?

Hypothetical impacts of state governments

The theoretical model developed in chapter 1 generates several hypotheses concerning the impact of state governments on ss air policy failures and possibilities. The first hypothesis of the model suggests that variance in state behavior is likely to be high in this policy area. While some matching between need and response may be apparent, with leading states thriving on the autonomy described above, overall state behavior is likely to be inconsistent. Second, the high level of interstate competition means that even the strongest state programs hesitate to exceed federal standards. Third, the low level of vertical involvement makes coordination across state lines, even between leading programs, difficult.

Overall State Behavior in Stationary Source Air Pollution Control

Evidence for overall state behavior in the propositions mentioned above is apparent in evaluations of SIPs, expenditures, enforcement procedures, and externalities.

State implementation plans

The first step in state procedures is the design of the State Implementation Plan (SIP). Considering the resources and technical expertise necessary but often unavailable to develop such plans, it is little wonder that numerous analysts have criticized this aspect of the law.[44] The history of SIP preparation and adherence suggests that state responsiveness has changed somewhat since passage of the CAA in 1970. Early SIPs were inconsistent at best. Evaluation of the most recent SIPs suggests fairly responsive state behavior.

The SIPs of the early 1970s were quite uneven. Business and industry interest groups recognized early that SIPs were important vehicles for SS policy. Because SIPs were supposed to propose specific procedures, the plans were, according to one study, "the primary target of attack from industry and other opponents of clean air regulations."[45] As a result of industry pressure, SIPs displayed considerable variance in regulations, compliance schedules, and exemptions from rules for emergencies such as accidents.[46] Most states hastily threw together a SIP and then allowed variances. Court challenges by environmentalists only led to more informal allowances by the states. One 1978 study concentrated on SIPs in nine states and concluded that they were notable for their "simplicity and their political sensitivity."[47] Another analysis argued that smelters were particularly successful in getting states to "propose SIP regulations weaker than those endorsed by the EPA" precisely because shutdowns would be so costly to the local economy.[48] Yet

Table 2.4 State Implementation Plan Rankings (dependent var. = 1 if strong, 0 if weak)

Independent Variables	SIP	Toxics
constant	−9.71*	−6.55
	(3.97)	(3.53)
aq[a]	−0.06*	−0.09
	(0.02)	(0.14)
mugsp[b]	0.25*	0.18*
	(0.08)	(0.06)
pipc[c]	0.23	0.23
	(0.20)	(0.19)
piindex[d]	2.32	−0.18
	(3.22)	(3.14)
fapc[e]	1.25	−0.37
	(3.46)	(2.94)
degrees of freedom	5	5
Chi square	25.4	21.28
sig level	>0.005	>0.005
prediction rate	84%	76%
*significant at the 0.05 level		

a. % of state populace living in excess SO_2 and particulates
b. manufacturing and utility sectors as % of g.s.p.
c. personal income per capita
d. index of voter turnout and party competition
e. federal aid dollars per capita

another analysis of the failure of states to develop effective SIPs concluded that such failures allowed "pollution havens" to exist.[49]

Many states have rewritten their SIPs since the time of these studies. Recent data on these plans suggest that states have become more responsive and realistic in their proposals. Table 2.4 displays testing of two measures for current SIPs. These measures are tested against the independent variables (described in chapter 1) that could affect state development of SIPs. Briefly, the pollution variable (*aq*) is a measure of percentage of state population living in areas with excess SO2 or particulates.[50] Economic pressure (*mugsp*) reflects percentage of state economy contributed by manufacturing and utilities. One might expect that these two measures (*aq* and *mugsp*) would be correlated but they are not, a fact which will be discussed later. Relative affluence (*pipc*) is operationalized by personal income per capita. Political culture (*piindex*) is measured by turnout and party competition. Finally, federal aid (*fapc*) uses dollars per capita contributed by the federal government through the 1970s for the control of air pollution.

Column 1 of table 2.4 utilizes a dichotomous dependent variable representing strong and weak SIPS. This differentiation was made on the basis of state-level data reflecting deficiencies in the SIP, EPA sanctions of the SIP, and inclusion of needed programs in the SIP.[51] The equation is statistically quite powerful as reflected in the strong Chi-square and the prediction rate of 84 percent. Air quality is negatively correlated, which we should expect since one of the components of the SIP variable is number of counties with deficient SIPs. However, the variable representing the portion of the state's economy contributed by manufacturing and utilities is also statistically quite powerful. This suggests that those states needing to develop strong SIPs have done so. The fact that states have strong polluting sectors does not prevent them from designing responsive plans.

Further evidence for this proposition is displayed in the second column of table 2.4. The dependent variable in this logit equation represents data concerning each state's air toxics program contained in the SIP. The states vary widely in the SIP specification of toxic air pollutant programs. As described earlier, the EPA has encouraged states to assume responsibility for the identification and regulation of toxics within their borders. As of mid-1987, roughly half of the states had developed such programs, but even these existing programs differed widely in terms of what pollutants were regulated, at what levels they were restricted, and how they were controlled. For example, the Connecticut program lists 853 toxic pollutants, whereas the Texas program has no specific list. New Jersey regulates dry cleaners as a source for toxics, but Texas specifically exempts them. Such basic concepts as technology required and ambient levels are subject to wide variation. The variance has become so wide, in fact, that prominent organizations of state and local air officials have called for federal establishment of minimum standards and oversight.[52] The second-column variable in table 2.4 differentiates states according to whether or not the toxic program utilizes ambient standards in its permit review, a very basic measure. Again, the equation is statistically strong and the pressure variable is quite powerful. In explaining the weak showing of the variable representing air quality, we should not expect it to be powerful since it is a measure of SO_2 and particulates, not toxics.

Evidence that states are developing responsive SIPS is interesting but not conclusive. After all, these are just plans with no guarantees for adherence or usage. Further, the measures of the plans used here do not account for the potential export of harmful pollutants. Still, states have developed some capacity to promulgate responsible guidelines.

Expenditures

Another possible measure of state performance is expenditures. States have considerable discretion over how much money is to be spent on SS air pollution control. Under Section 105(b), states are required to use federal money for supplementing, not supplanting, their own air program expenditures. Further, grantees are not entitled to federal funds in subsequent years wherein their own funding levels do not exceed previous years. EPA can allow exceptions for this if all state programs are reduced. In a survey of three regions, the General Accounting Office (GAO) found that the federal government did not utilize procedures to ensure that states were maintaining previous levels of effort. For one thing, the determination that funds were not supplanted is quite subjective. Additionally, the GAO cited numerous instances when, in fact, levels of effort by grantees had not been maintained.[53] State discretion is thus enhanced by low federal oversight.

Analysis of one year of state spending is presented in table 2.5. The

Table 2.5 Determinants of State Air Expenditures (dependent var. = air expenditures per cap in FY86)

Independent Variables	With Federal Aid	Omit Federal Aid
constant	8.58	1.15
	(1.22)	(1.15)
aq[a]	0.15*	0.14*
	(0.06)	(0.05)
mugsp[b]	−0.36	−0.40
	(0.21)	(0.21)
pipc[c]	0.25	0.26
	(0.69)	(0.69)
piindex[d]	2.57	2.91*
	(1.32)	(1.24)
fapc[e]	0.86	—
	(1.11)	—
R square	0.34	0.33
Adjusted R square	0.26	0.27
*significant at the 0.05 level		

a. % of state populace living in excess SO_2 or particulates
b. manufacturing and utilities sectors as % of g.s.p.
c. personal income per capita
d. index of party competition and turnout
e. dollars of federal aid per capita

Table 2.6 Coefficients of Variation in Public Expenditures

	State Air	State Water	State Water*
1973–74	0.71	1.09	0.61
1975–76	1.11	1.14	0.54
1977–78	1.45	1.16	0.54
1979–80	1.42	1.09	0.44

*water spending includes the waste treatment grants program

dependent variable in the table is state expenditure per capita for air pollution control in fiscal year 1986. The first column includes the independent variable operationalizing federal aid. Because the federal dollars are conceivably substituted for state dollars, it is difficult to separate these effects. Thus the same equation was run without the federal aid variable and is presented in the second column of table 2.5. Results in both columns suggest that state spending is significantly and positively affected by the lack of air quality (*aq*) in the state and by the interest of the electorate (*piindex*). The matching of expenditures to need is thus partially supported.

Expenditure data can also be utilized to display the high variance that will occur with considerable state discretion. Table 2.6 displays a comparison between ss air policy and more federally controlled point source (factories and utilities) water spending. Water policy expenditures are reflected by two measures, one including the massive waste treatment grant program (described in chapter 3) and one without. The coefficient of variation (standard deviation/mean) has been calculated for each year of each measure of state expenditures.[54] The table displays average coefficients for consecutive years, combining years to avoid idiosyncracies. The comparison is revealing. As state policymakers recognized their independence and developed their own means of control, the variance in air spending surpassed that of both measures of water spending. The low variance displayed by the water measure that includes the grant program is predictable considering the inclusion of matching grants. The important point is that variance in this simple measure is indeed higher in ss air policy than in the water counterpart.

Although state expenditure has been used to measure variations in resource utilization for many policies, such operationalizations are subject to some question because some states may be more efficient in the usage of their funds than others are.[55] Measuring policy expenditure on a per capita basis does not solve the efficiency question but it does provide a systematic means of comparing priorities between states.[56] Examination of one year's data indicates a positive correlation between high presence of air pollution and high prioritization of air control expenditures.

Enforcement procedures

Once SIPS are designed and funds appropriated, the effectiveness of state programs depends on enforcement. Enforcement includes regulations, inspections, penalties, and sanctions. Because these specific aspects are determined at the subnational level, high variance exists between states, thus making overall assessments problematic.

State enforcement efforts have varied widely in terms of measurement and regulations. For example, at one time SO_2 emissions were regulated on the basis of heat input (lb SO_2/mill BTU) by thirty-six state and local agencies, while twenty-eight agencies restricted the amount of sulfur content in the fuel. Measurement differences are further complicated by exceptions to regulations that vary, such as when minutes per hour standards can be violated for equipment malfunctions and maintenance.[57] Thus objective assessments of state enforcement efforts are rare.

Official reports of state enforcement activities are often quite positive. As early as 1977, the EPA reported that 90 percent of stationary sources of air pollution were in compliance, but this referred to the installation of equipment, not the reduction of discharge.[58] A later EPA audit of air policy implementation by the states retains the encouraging tone but admits that "their overall score relative to actually meeting all the literal objectives of the CAA shows room for improvement."[59]

Other studies are more discerning. A recent GAO study describes that while 95 percent of the emitting stationary sources were inspected during FY1984, 39 percent of those inspections were done inadequately.[60] Further, these efforts largely rely on voluntary self-monitoring by polluters since continual vigilance by authorities would be costly and impractical.[61] Not only are inspections announced in advance, thus allowing polluters to prepare, but even when penalties are imposed by state authorities, they are often very small.[62] Finally, state administrators negotiate with violators to set schedules for compliant behavior. Many of those sources considered on schedule will not be fully compliant for many years.[63]

Combining appraisals of enforcement statistics provides a measure of state performance. The Fund for Renewable Energy and the Environment has compiled statistics on deficiencies, numbers of monitors, inspections, and violations and used them to rank state programs on a 1 to 10 scale. That rank is presented as a dependent variable in table 2.7. The variable is tested (with ordinary least squares) against an equation representing the independent variables described above. The results are quite robust, with several independent variables statistically significant. Since the dependent variable represents a finite number of categories, one should be careful of these regression

Table 2.7   State Air Pollution Control Program Rankings (dependent var. =
ranking of state program: 1–10, 0 or 1)

Independent Variables	Rank (1–10)	Rank (0 or 1)
constant	−8.12*	−12.45*
	(2.38)	(4.36)
aq[a]	0.13	0.04
	(0.11)	(0.16)
mugsp[b]	0.14*	0.16*
	(0.04)	(0.06)
pipc[c]	0.74*	0.74*
	(0.14)	(0.26)
piindex[d]	4.27	6.04
	(2.59)	(5.75)
fapc[e]	−0.62*	−0.91*
	(0.22)	(0.40)
R square	0.56	
Adjusted R square	0.51	
Chi square		26.02
sig level		>0.005
prediction rate		74%
*significant at the 0.05 level		

a. % of state populace living in excess SO_2 or particulates
b. manufacturing and utilities sectors as % of g.s.p.
c. personal income per capita
d. index of party competition and turnout
e. federal aid per capita in dollars

results. To see if the numbers are skewed, the ranking has been broken down
in the second column of table 2.7 into a simple dichotomous variable representing strong and weak air pollution control programs and run in a logit
equation. Again, the equation is significant. Further, the presence of polluters
and the relative affluence of the population are significant determinants of
strong enforcement programs. The federal aid variable is negative, suggesting that federal aid has not determined the existence of strong state enforcement programs.

These results suggest several things, but they should not be interpreted as
definitive. First, they suggest the appropriateness of these particular independent variables for predicting state performance. In the case of ss air
policy, the results are consistent with the characterization of federal influence
as low and the development of strong state programs as dependent on
indigenous state factors. Since they do depend on indigenous factors, vari-

ance between state programs can be quite high. Second, they provide prima facie evidence for state responsiveness, as those states with the largest presence of polluters are the most likely to respond, especially if they enjoy abundant resources and dynamic institutions. Third, however, these results can only be taken so far. The rankings of state programs in these tables do not include any measure for how much of the pollution is being externalized to other states. Again, the nature of air pollution and the structure of the federal response allows states to have what appear to be strong control programs but which may also rely to some extent on exporting the problem.

Acid rain

Potential exportation of air pollution is apparent in the case of acid rain. This case also provides evidence for the hypothesized lack of interstate coordination and state reluctance to supersede federal guidelines in this policy area. States can meet federal standards by externalizing ss discharge in order to achieve quality goals without reducing indigenous industrial production. Localized ambient quality goals create incentives for states to export pollution problems when possible. While the states can either coordinate or separately exceed federal directives to address this problem, the influence of economic interest groups within polluting states has made such behavior, for the most part, unlikely.

Acid rain results when sulfur dioxide and nitrogen oxide emissions combine with oxygen in the atmosphere to create sulfuric and nitric acids. These acids can then be transported by wind currents to precipitate in other locations. Since ambient air quality is measured in finite areas, the achievement of attainment targets by exporting pollutants may be a tempting prospect for those states with heavy industrial presence and powerful local interests. The indigenous factor dominating state behavior on this issue is the presence of coal-burning plants that produce sulfur dioxides. Abundant particularly in the Midwest, these industries compel states to foster behavior allowing them to meet externally imposed targets without curtailing internal economic activity. Studies have shown that nearly half of the sulfate levels in the Adirondack Mountains of New York originated in midwestern states.[64] Acid rain is not limited to the Northeast. Recent studies have found acid deposition in Colorado, Idaho, and New Mexico.[65]

The federal structure in air policy creates incentives for states to export pollution. Standards are designed and enforced on the basis of local, ground-level concentrations. Most current methods measure pollutants for only relatively short distances, often no more than fifty kilometers.[66] Polluters can thus avoid penalties, and states can encourage such behavior with their own sanctions, through techniques of dispersion enhancement. Operating times, for example, can be staggered to pollute when meteorological conditions are conducive to dispersal. Tall smokestacks can be built so that emissions are dispersed by higher altitude winds. More than 90 percent of SO_2 emissions in North America in 1978–79 emanated from stacks higher than 325 feet and 60 percent came from stacks of 650 feet and higher.[67] More significantly, after ten years of federal intervention, the number of stacks of 650 feet and higher increased from less than a dozen (in 1969) to 180 (by 1979).[68] The average height of smokestacks in the United States has jumped from 243 feet in 1960 to 730 feet in 1980. This dramatic increase should not be considered coincidental.

State leadership in this policy area is thus potentially negated by self-interest. The self-interested behavior of states in this matter is best illustrated by several anecdotes. The SIP for the state of Georgia was thrown out of court in 1974 because it relied "almost entirely" on tall stacks to meet ambient standards.[69] SIPs in Ohio, Illinois, and Indiana were all severely affected by litigations from polluting industries.[70] That a state can be on both sides of this issue is shown by New York's involvement in two litigations. In the first major set of cases on this issue (*Connecticut v.* EPA, 696 F.2d 147 [1982]; *Connecticut Fund for the Environment v.* EPA, 696 F.2d 169 [1982]; and *Connecticut Fund for the Environment v.* EPA, 696 F.2d 179 [1982]), New York was accused by Connecticut of granting variances (exemptions) to polluters, which affected (by downwind transfer) air quality in the latter state. The Supreme Court's ruling that such variances were illegal only if they consumed most of the offended state's clean air margin "places states in competition with upwind states for the use of the clean air margin."[71] The second litigation involved a series of suits by northeastern states, including New York, against seven midwestern states on the grounds that transmitted externalities prevent local attainment of ambient quality standards.[72]

Still, some state governments have responded to acid rain problems. The New York situation illustrates the air pollution version of "where you stand depends on where you sit." State governments, particularly those "sitting" downwind of industrial pollution, can attempt to address the situation internally. New York passed legislation in 1984 calling for a 30 percent reduction in SO_2 emissions by the mid-1990s. New Hampshire followed suit in 1985 by stipulating a 50 percent reduction in SO_2 over a ten-year period. Minnesota, Wisconsin, and Massachusetts all have their own acid rain control laws. Other states conduct research on acid rain.

Table 2.8 represents two attempts to systematically consider state redress of acid rain. The first column uses a simple dichotomous variable to examine whether or not the state conducts research on acid rain. Although the equa-

Table 2.8 Determinants of Acid Rain Exports (dependent var. = researches acid rain; % SO_2 reduction)

Independent Variables	Research	SO_2 Reduction
constant	−4.01	−4.34
	(2.81)	(2.94)
aq[a]	0.04	0.17
	(0.13)	(0.13)
mugsp[b]	0.52	2.10*
	(0.49)	(0.51)
pipc[c]	0.23	0.11
	(0.16)	(0.17)
piindex[d]	8.49	3.56
	(5.28)	(3.19)
fapc[e]	−0.52	−0.71*
	(0.29)	(0.27)
Chi square	11.07	
sig level	>0.05	
prediction rate	72%	
R square		0.44
Adjusted R square		0.37
*significant at the 0.05 level		

a. % of state populace living in excess SO_2 or particulates
b. manufacturing and utilities sectors as % of g.s.p.
c. personal income per capita
d. index of party competition and turnout
e. dollars of federal aid per capita

tion is fairly strong, none of the independent variables are statistically significant. This could reflect, in part, the fact that many states do not export SO2 or nitrous oxide and thus have no need to research acid rain. It could also reflect the possibility that state programs are not as responsive to this form of pollution as the previous analyses might have led us to predict.

The second column in table 2.8 provides some evidence for this latter assertion. The dependent variable in this table is measured in terms of acid rain contribution by state. Data for this measure are taken from congressional formulation of the Waxman-Sikorski bill (H.R. 3400) of 1984. This bill was the most comprehensive attempt in Congress to abate acid rain before the current Bush proposals. The bill identified states that needed to reduce sulfur dioxide emissions and the necessary amounts of reduction.[73] The numbers in this column represent the percentage of the national total constituted by each state's proposed SO2 reduction. The equation is fairly powerful, but what is more important is the significance of the measure for presence of manufacturing and utility sectors. Obviously, states with heavy presence of these sectors are the ones that would be forced to reduce their emissions. Since the bill was directed at acid rain, one can argue that those states are the ones externalizing their pollutants.

The 1990 revisions to the Clean Air Act reflect an increased federal involvement in state decisions concerning acid rain. The legislation requires reduction of sulfur dioxide emissions to 1.2 pounds per million BTUs. This will necessitate removing ten million tons of sulfur dioxide emissions in the United States by 2000. Many midwestern coal-burning power plants will need to make nearly 50 percent cuts in sulfur dioxide emissions. Utilities are expected to add more scrubbers (roughly 150 are now in usage), switch to low-sulfur coal, and raise consumer bills (by as much as 25 percent) to pay for the changes.[74] Federal specification of offending utilities and mandated reductions are manifestations of exhausted patience with state inertia on this issue.

Overall performance of the states

State leadership in this policy area is inconsistent. Matching between need and response is quite variable. States are more likely to have strong programs if the problem is large, the resources are abundant, and the political institutions at the state level are vigorous. States vary widely on these factors. However, the stronger the economic interest groups, the more likely that states will rely on the export of their pollution problems so that they can still meet ambient goals without threatening their most valuable constituents. Thus the presence of economic pressure also makes interstate coordination and state supersedure of federal standards rare.

Leading State Behavior in Stationary Source Pollution Control

The dimensions of federalism strongly affect the leadership efforts of state governments in ss air policy. Individual states may develop exemplary policies, but even the best programs find it difficult to exceed federal guidelines and to overcome interstate tensions. These phenomena are illustrated by description and explanation of one leading state and its program: Wisconsin. The preceding analysis identified Wisconsin as one of the leading states in ss air pollution control. Wisconsin displays a strong sip, conducts research on acid rain, has a developed toxics program, and is scored by outside authorities (on the scale used in table 2.7) as 9 out of 10. I emphasize that the following discussion concerns one of the strongest programs and that such behavior is not typical of all state governments. In fact, the difficulties faced by Wisconsin are, in all likelihood, magnified in other states.

Description

Wisconsin's political behavior is best characterized as unpredictable. Since the three-party days of Robert LaFollette's Progressives, the state has often voted counter to national trends in presidential elections.[75] The state has been the home of politicians as diverse as Joe McCarthy and William Proxmire. Contrasts continue to make the state's behavior in air pollution control seem unpredictable. The percentage of gross state product contributed by manufacturing and utilities (35.59) is the seventh highest in the country.[76] According to the proposed requirements of H.R. 3400 (the unsuccessful 1980s acid rain bill), Wisconsin's share of national SO_2 emission reductions (6.89 percent) would have been higher than all states except Indiana and Ohio.[77] Despite this obvious industrial presence, Wisconsin has one of the three strongest ss air pollution control programs in the country. This case study describes the program and then suggests some reasons for Wisconsin's behavior.

Wisconsin's environmental program, in general, is "comprehensive and exacting."[78] In the inaugural rating of state programs, FREE ranked Wisconsin the highest overall of all states in their environmental efforts. The state "showed a consistent high level of program strength—demonstrating broad commitment to the environment and public health."[79] State policymakers and politicians are proud of that legacy. The original Earth Day was conceived of by a Wisconsin senator, Gaylord Nelson.

While the entire environmental effort is cited, the air pollution control component in particular is emphasized. FREE ranked the Wisconsin air program behind only California.[80] California officials provided some external validity for this ranking by citing Wisconsin and New Jersey as other states with strong programs. Wisconsin administrators speculated that California was ranked higher only because the FREE ranking included both mobile and stationary sources. They were confident that their ss air control efforts took a back seat to no state. This section describes Wisconsin's air efforts in terms of regulation, funding, and education.

Regulation is the responsibility of offices within the Division of Environmental Standards. This division is one part of the Department of Natural Resources (DNR). The DNR is relatively insulated from political influence. Its leadership consists of a seven-member board whose members are appointed for staggered terms. The secretary of the DNR can be removed only by the board and not by elected politicians, such as the governor. As one official suggested to me, this insulation is not likely to change since politicians can now pass tough decisions along to the DNR and claim reliance on nonpartisan expertise.

Wisconsin's regulatory efforts center around permits for polluters. The process works as follows. Rules are proposed by the DNR concerning specific areas of pollution, such as sulfur emissions, particulates, and organic compounds. Following public hearings, the rule is then subject to approval by the independent DNR board. If approved (the norm), the state legislature has thirty days to veto the rule. Without veto, the rules become part of the Wisconsin SIP. The rules define pollutants, specify emission limits for various polluting processes, and explicate compliance schedules and variances. For example, Chapter 415 of the SIP defines particulates such as fugitive dust, specifies emission limits in forms such as 0.2 pounds of particulate matter per 1,000 pounds of gas for cement kilns, and then mandates compliance plans be submitted by 1 July 1988.[81] Following promulgation of the rule, each significant polluter, such as an individual cement kiln, must obtain permits for operation. Wisconsin has "unusually pervasive regulations for larger facilities" and requires all permits be in hand before construction of a new polluting source can begin.[82]

One major attribute of Wisconsin's program is speed in this process. State administrators boast of the speed with which they can get a rule approved so that permits can be promulgated. According to one official, this is done substantially faster than in other states, where permittees can tie up rules for years with protests. If not controversial, the rule can be proposed, hearings held, and board approval obtained within nine months in Wisconsin. Administrators are also quite proud of the timeliness of permit processing. The state guide for industrial compliance suggests that, when permittees are forthcoming with information, the process can usually be completed within ninety days.[83] In the foreword to her comparative guide to state programs, Jessup cites a recent survey that shows even the regulated community applauds Wisconsin regulators for their efficiency and knowledge.[84] Speed enables rapid responsiveness by the state to air pollution problems.

The permits exist in many forms. Wisconsin has authority to assign permits in all of the major programs: National Emissions Standards for Hazardous Air Pollutants (NESHAPS), NSPS, and PSD. The state is supposed to use whichever is the most restrictive of those that apply. The overlap makes the programs complex and confusing, but state officials are confident that they can handle the complexity even without federal guidance. Federal oversight is close enough in the PSD area that EPA administrators can tell the state when a permit is too lenient. This has rarely happened. Instead, Wisconsin often sets the standard in many control areas. Nevertheless, one state official told me that they do not mind having the federal "gorilla in the closet," which can be turned loose if polluters object too much. Most of the time, the state enjoys considerable discretion in implementation of all these aspects of air pollution control.

Compliance is a major concern of state officials. Administrators stated that most polluters cooperate on a voluntary basis. The DNR facilitates such cooperation by providing thorough and explicit guidelines for compliant behavior.[85] Agency officials do admit that pressure is present, particularly when rules such as the "toxic rule" are being considered. Industries do threaten to move and some policymakers admit that loss of the industrial base is a concern. However, the impact of relocation threats is moderated by two factors. First, the DNR involves polluters in the requirement-setting process. This involvement is appreciated by the regulatees. The more it is present, the less likely threats will be voiced, let alone carried out. Second, officials realize that relocation decisions are often determined by factors other than state policies, such as markets, resources, and cheap labor. One administrator used the example of paper mills moving south because of the abundance of trees. The more that policymakers recognize the relative importance of other factors, the less impact voiced threats have on the policy at hand.

Funding for the air pollution control program is adequate and substantially self-generating. In fiscal year 1986, for example, Wisconsin spent nearly $5 million on its air program. In per capita terms, this is in the middle range of states.[86] Roughly 40 percent of the state's funds come from federal grants. Of the rest, approximately 25 percent is generated through permit fees that come directly to the DNR and another 25 percent derives from emission fees that are paid to the state and then funneled back to the agency. The permit fees for the construction of major and minor sources are, respectively, $4,500 and $1,300. The permit fees for modification of major and minor sources are $3,000 and $1,000.[87] Annual fees are also charged to defray costs of inspections and monitoring. Operators of sources that emit more than 0.25 tons per day or 50 tons per year of certain pollutants must submit emission fees along with annual reports to the state legislature.[88]

The educational component of Wisconsin's pollution control efforts is directed at future as well as at present constituents. The state has required instruction in conservation and natural resources in grade schools since the days of Aldo Leopold in the 1930s. The heart of the existing program is a curriculum planning requirement for grades K through 12, which is one of the numerous standards that school districts must meet in order to obtain funding. The requirement, created in 1985, mandates that environmental education should be integrated into all programs, with special emphasis in certain areas such as the sciences. If districts fail to develop plans, the state can withhold up to 25 percent of their financial aid. Although several states have environmental emphases in their grade schools, one official in the Wisconsin program asserted that his state was unique in certain aspects, such as the requirement of environmental competency for teacher certification in certain areas. The Department of Public Instruction also cooperates with the private, nonprofit organization Wisconsin Association for Environmental Education, Inc., to distribute newsletters and coordinate adult programs.[89]

Leadership

Leadership in the Wisconsin SS air program is manifest in experimentation with market incentives, the state's acid rain law, and its toxics program. The state does utilize economic incentives such as banking and offsets, particularly in the southeastern part of the state, which is not only industrial but also receives pollution from the Chicago area. State administrators have expressed a willingness to use such methods but also a wariness to complete reliance on them. As one official said, "bubbles have been only a marginal success so far because the progress probably would've been made anyway." The DNR is interested in the experiences of other states in such programs, although such information is not easily obtained. In fact, several administrators suggested that the federal government could provide more service by acting as a clearinghouse for such information. However, the relatively low federal involvement in this area has so far precluded extensive interstate coordination.

The state of Wisconsin is particularly proud of its behavior in the area of acid rain. During the mid-1980s, the state was one of the first to pass an acid rain law that capped SO_2 emissions. Limitations on sulfur emissions are specified in great detail in Chapter NR 417 of the SIP.[90] Detail is provided on types of fuels, types of sources, and compliance schedules. In addition to controlling its own contributions to acid rain, Wisconsin is contesting pollution from outside its borders. The state initiated a suit against the EPA in 1987 to compel the federal agency to force Illinois and Indiana to change their polluting behavior. The suit was based on the nonattainment status of southeastern Wisconsin due to pollution from southern neighbors, which prevented industrial development in that area. The state won its suit in January 1989. However, the court-ordered solution was not as strong as Wisconsin wanted, so officials negotiated a settlement with Illinois for more prompt and effective behavior. The settlement calls for interim controls, realistic models, and the matching of Illinois and Indiana programs to those of Wisconsin in certain areas. One offical cited another regional study now under way that will eventually promulgate control strategies for the entire region, including any effects Wisconsin's pollution is having on Michigan's air. As this official concluded, the states are "dragging" the federal government into what many see as necessary regional approaches to the acid rain problem. This should not be overstated. The state is willing to fill in gaps in federal air control policy but less willing to supersede federal standards that do exist. I asked two separate officials if the state had ever surpassed federal requirements and both replied "never." Further, the reluctant behavior of Indiana and Illinois in this issue supports the hypothesis that interstate coordination may have to be forced rather than voluntary.

Finally, the passage of Wisconsin's "toxics rule" is an interesting and illustrative story. The rule resulted from the efforts of the Air Toxics Task Force. The ATTF was formed in 1983 to formulate a rule governing hazardous substances not covered by the CAA. The ATTF consisted of two industry representatives, two environmental representatives, an epidemiologist, and representatives from the state departments of development and health. The head of the ATTF, James Rickun, was an employee of the DNR. As Rickun describes, the process was heated and intense. The ATTF met at least monthly for two and one-half years before promulgating a recommendation. The ATTF recommendation was passed by a 5-to-2 vote, with the environmentalists casting the negatives. According to Rickun, these no votes were political decisions based on fears that if the rule was perceived as an environmentalists' rule, it would be chipped away in upcoming hearings. The recommendation was then considered in a series of six public hearings during 1986. During the hearings process, industry withdrew its support despite a survey by DNR and the chamber of commerce that showed that compliance costs would not be exorbitant. Industry representatives claimed that the rule was more stringent than necessary and tougher than those of other states. If passed, these spokespersons argued, industrial relocation would not be out of the question. Nevertheless, the DNR board approved the ATTF recommendation and the state legislature neither vetoed nor changed it substantially. The rule is now being contested in court by a strong coalition of industrial groups on the grounds that the state exceeded its statutory authority.[91] Again, the potential for horizontal competition makes supersedure of specified standards problematic.

The Wisconsin toxics rule is thorough and explicit. It specifies limits, sources, and compliance schedules.[92] Rickun himself does not believe that it is too stringent. Rather, he suggests several reasons why it is praiseworthy. First, it is reasonable in that it pioneers the technology-based, rather than risk-based, approach to carcinogens that is now being used in the latest version of the CAA. Thus instead of forcing industry to meet one-in-a-million assessments, industry must reduce emissions to the best of technological ability. Second, the rule is comprehensive and utilizes a considerable body of research on carcinogens already completed. Third, the rule addresses existing sources of pollution rather than taking the politically expeditious route of setting limits only on new sources. Still, Wisconsin's difficulty (as apparent in Rickun's resignation) in passing their toxics rule suggests that passage in less receptive states will be problematic.

Explanation

Environmental awareness is pervasive in Wisconsin. Upon walking into the DNR office building, I was confronted with dozens of pieces of imaginative art offered by the state's schoolchildren. The current theme was Earth Day, so pictures abounded of pastoral scenes being affected by industrial pollution. One particular drawing caught my eye, of a skunk, downwind from a factory, with a clothespin over its nose. Someone joked that the factory must have been in Chicago. Several reasons explain Wisconsin's behavior in air pollution control.

First, the state recognizes the need for serious response. Although not as heavily industrial as some of its neighbors, the state does have some 1,200 sources under permit. The state may be famous for pastoral scenes and dairy farms, but it also produces beer, paper, clothes, and machinery. Recognition of pollution problems has existed for decades, but awareness and urgency have been even more acute in the last eight or nine years, according to people in state government. During the same period, officials have gained the maturity and expertise to develop comprehensive programs. As one high official said, the state used to look to the EPA for guidance but now state administrators have more knowledge and experience than do their federal counterparts.

Second, the state possesses the resources to afford strong programs. Even though Wisconsin, as described earlier, does not spend more than many other states, the funds are there when needed. Only seven states ranked higher in state government tax revenue per capita in 1983 than Wisconsin.[93] As described before, tax revenue is only one of the sources of funding for the air pollution control program. The state is thorough and efficient in the collection of permit and emission fees. This self-generation of funds provides support and autonomy for the DNR. The head of the permits section stated that one of his goals was to become even more independently financed in the future.

Some degree of autonomy is already present in the political institutions of the state, the third factor contributing to Wisconsin's effective program. The relative insularity of the DNR board has already been described. Since the secretary of the DNR reports only to the board, he or she can perform the duties of the job without fear of political interference. The governor does appoint the board, but meddling with its members can be politically dangerous. The DNR itself has enough support within the environmental community and enough reputation within the business community that it is relatively insulated from excessive outside pressures. Officials described environmental groups as providing strong counterpressure to industrial lobbying. Further, the agency is relatively insulated from federal influence. Officials cited few unwanted instances of federal meddling in their activities.

Fourth and perhaps most important, the political culture of the state of Wisconsin is strongly oriented toward environmental awareness. The state is blessed with abundant natural resources and a progressive spirit. In general, people are active outdoors and also quite learned and politically aware. As a result, they are involved in environmental efforts and informed about environmental issues. Parties compete for environmental attention. What was once the host for a three-party system is now a state with competitive "two-party, liberal-conservative politics."[94] The state is also home to a thriving tourist industry. Thus, for economic reasons as well, many citizens adopt environmental ideals and encourage the preservation of the state's natural beauty.

Conclusions

The dimensions of federalism governing stationary source air pollution control policy can be characterized as low in vertical involvement and high in horizontal competition. Designation of vertical involvement as low is based on the delegation of means of attainment to the state level. The discretion created by such delegation fosters relatively autonomous behavior by state governments. High potential for interstate competition is created by the nature of the problem. Stationary source air polluters are mobile enough and large enough as single entities to threaten states with relocation. Further, the pollutants themselves can easily travel over state lines, especially when released at high levels of the atmosphere.

The effects of these dimensions on state leadership are apparent in evidence supporting the three hypotheses of chapter 1.

Hypothesis 1. The fifty state governments, as the model predicts, exhibit wide variance in ss air pollution control behavior. Some correlation exists between presence of the problem and state responsiveness, although the matching is more direct for stated intentions (table 2.4) and expenditures (table 2.5) than it is for behavior such as enforcement (table 2.7) and acid rain control (table 2.8) that might be more objectionable to each state's economic sector.

Hypothesis 2. High interstate competition makes supersedure of federal guidelines unlikely in most cases and delayed even for the strongest states. Even in leading state programs such as Wisconsin's, adoption of market incentives has been slow and passage of toxics regulations has been painful and protracted.

Hypothesis 3. The lack of extensive vertical involvement is most apparent in the persistence of acid rain problems. State leaders may develop procedures to abate the export of harmful pollutants, as did Wisconsin, but attaining the cooperation of neighboring states can require legal action.

During the last twenty years there has been some progress in the control of stationary source air pollution, but serious problems have also persisted. The effects of the dimensions of federalism on state leadership suggest that state governments alone cannot resolve some of these issues. Indeed, policymakers need to reconsider just when it becomes necessary to release the "federal gorilla" from the closet. The gorilla's first task should be to coordinate state efforts to control acid rain. When President Bush signed the 1990 Clean Air Act on 15 November, he at least opened the closet door.

3   Point Source Water Pollution Control

In the spring of 1985 Gary Hunt and the other members of the North Carolina Pollution Prevention Program hosted a conference. Fifteen administrators from five other pioneering state prevention programs gathered in Raleigh to discuss their fledgling efforts at an aspect of pollution control unaddressed by federal legislation. Five years later, the conference is an annual event partially sponsored by the federal EPA with representatives from over thirty-five states attending. Pollution prevention is an integral part of North Carolina's point source (PS) water pollution control program. The development and dissemination of such efforts reflects the lead that states have taken in this policy area.

This chapter examines the role of the states in this policy area, discusses their overall behavior, and illustrates leading efforts with a description of North Carolina's programs. The above story foreshadows the broader findings in the chapter. First, state leaders willingly fill in gaps in the federal policy structure while not exceeding existing federal guidelines. Second, progressive efforts are readily disseminated across state lines.

The Role of the States in Point Source
Water Pollution Control

A comparison of the institutions created at the federal level for control of point source (PS) water and air pollution is interesting. Point sources of water pollution are similar to stationary sources of air pollution in that discharge emanates from fixed, specific origins such as factories and municipal water treatment plants. The 1972 Federal Water Pollution Control Act (FWPCA) utilizes the states in a means-forcing system rather than in the ends-forcing system relied upon by stationary source air policies. Specifically, the federal government emphasizes means of attainment rather than ambient quality goals. Although modified somewhat since, this basic difference between the policies has been maintained. Even today, state PS water programs are subject to greater federal intervention than are stationary source air programs. While different in this vertical dimension, PS water and air policies are similar in the high level of interstate interaction. Polluters and pollution have the potential to cross state lines. Both federal intervention and interstate interaction are discussed more fully below.

High vertical involvement

The high degree of federal involvement in state PS water pollution control programs has existed since 1972. It was inspired by the demands of environmentalists and has been guaranteed by the tremendous amount of grant money distributed to the states. High levels of funding from the federal pork barrel motivate national policymakers to maintain significant mechanisms of intervention into state behavior in this policy.

Vertical involvement in water pollution control was more potential than reality until the 1970s. The federal role dates to the Refuse Act of 1899, which prohibited industrial waste discharge without a permit from the U.S. Army Corps of Engineers. However, despite the fact that industry and agriculture have been responsible for roughly 90 percent of United States water withdrawals for decades, and thus presumably have created waste discharges, the law was not utilized until the 1970s.[1] In the years preceding 1972, grant allocations and other legislation seemed to increase the federal role in enforcement issues, but the scant efforts that did occur were not directed at the areas of the greatest need.[2] Analysts at the start of the Environmental Decade observed that: "Pollution is welded to political power. . . . Wherever water is bad the industries and cities that made it that way are ready to use their influence and economic leverage on every level of government to stave off abatement."[3] As a result of government inaction, pollution cesspools developed. The Cuyahoga River, for example, was so littered with trash that it actually caught on fire in 1969.

State and local governments were not compensating for federal inertia. One early study reported that state programs were usually run "competently" and some with "distinction," but these conclusions were based on a survey of six states that were admittedly picked because they had "relatively strong programs."[4] Only half of the states had set standards by 1970, and procedures for forcing compliance were slow and conciliatory.[5] Testimony in 1971 Senate hearings contended that, because of understaffing and low budgets, less than twenty states were in fair shape on water pollution control.[6] Nor, as developing evidence would show, did regional solutions seem any more promising than individual state efforts were.[7]

The 1972 FWPCA legislated several major changes in the control of water pollution in the United States. With a stated goal of eliminating discharges by 1985, the legislation limited effluents by setting standards on entire classes of industry regardless of location. The National Pollution Discharge Elimination System (NPDES) requires industrial and municipal treatment facilities to obtain permits that limit their discharge by type and amount. Section 301(a) mandates that every PS polluter must obtain a discharge permit from either the EPA or the authorized state agency. The FWPCA set deadlines for industries to install "best practicable" technology (1977) and ultimately "best available" equipment (1983).[8] The law mandated municipal publicly owned treatment works (POTWs) to provide secondary treatment of water by 1977 and best practicable treatment by 1983. Secondary treatment involves the use of bacteria and other biological methods to remove wastewater solids, which are disposed of as sludge. It is more stringent than primary treatment (gravity used to remove sludge) but not as developed as advanced treatment, which uses chemicals. The bill also legalized citizen suits against polluters and federal agencies if their economic or recreational interests were adversely affected by pollution. State governments were to implement, but not modify, these limits and procedures.

The FWPCA does not emphasize water quality goals. National targets are not more specific than the achievement of "fishable-swimmable" waters by 1983. The indicators described later in table 3.2 give some sense of measures that could have been used for national ambient standards. The legislation makes localized targets possible in the use of ambient quality targets as well as the technology-forcing effluent limits on specific bodies of water. Technically, water quality levels are to be utilized when technology-based limits are deemed insufficient to enable bodies of water to support their designated uses (Sec. 303-d). For example, waters categorized as fishable are supposed to be able to support a certain level of fishlife. States are then to establish "total maximum daily loads," stringent limits based on the maximum amount of pollutants waters can receive without violating quality standards.

This quality approach, the water equivalent of ambient air standards, relies on mathematical models that enable determination of wasteload allocations (WLAS) for individual polluters of the bodies of water. Such an approach immediately differs from national ambient standards in two important ways. First, the states determine the final usage for a body of water, thereby incurring widely varying sets of standards. Second, modeling was, for a long time, unsophisticated and idiosyncratic. Usage, as a result, was rare.[9] Nevertheless, it is important to keep in mind that Section 510 allows states to use effluent limits or quality targets and to use whichever is more stringent if they so desire.

The FWPCA included one of the largest public works projects ever legislated by Congress. The bill contained authorizations of more than $24 billion, mostly for construction of waste treatment plants, an amount that led President Nixon to attempt in vain to veto the measure. Although the exact distribution is somewhat complex, the conference compromise based allocation more on congressional need than on pollution.[10] Briefly, all states are entitled to at least one-half of 1 percent of the federal funds. This universalism guaranteed a unanimous vote on passage of the bill and override of the veto. Not surprisingly then, this aspect and not the intergovernmental implications of the legislation received the most attention by analysts and the public.[11] The grant money also ensured the lack of unified opposition by state policymakers to serious federal intervention in water issues.

The high level of pork available in the PS water policy barrel has fostered two other developments affecting the role of the states. The first impact is an orientation around process rather than results. Grants for plant operations are directed at specific projects. This emphasis on means of attainment rather than on whole programs (as in air) reinforces the PS water orientation around methods instead of overall behavior.[12] Even when planning costs, rather than specific actions, are eligible for federal funds as in Section 208 of the legislation, the emphasis is on projects rather than on results, on means rather than on ends.[13]

The second impact of the waste treatment grants program is the prevention of significant diminution of federal involvement over the years. The federal share of the construction costs for POTWs was initially at least 75 percent. Annual appropriations were somewhat uneven, especially when President Nixon impounded half of the initial allocation.[14] Nevertheless, the money ensured federal involvement in every stage of POTW planning and construction.[15] While the share of federal money has been changed over the years, the level has remained high enough to ensure high national attention to state behavior.

The 1977 amendments to the FWPCA modified the state role in PS water policy. By then, nearly 85 percent of major PS dischargers were in compliance with BPT standards, but only 30 percent of major municipalities had achieved secondary treatment.[16] Congress responded to the latter by requiring states to set aside 3 percent of their POTW allotment for innovative treatment technology. Within four years, more than 800 municipal projects were using new or alternative technologies.[17] Two of the 1977 amendments were stimulated by court decisions. The most controversial extended federal Corps of Engineers jurisdiction over project permits to most navigable waterways and adjacent wetlands.[18] The second institutionalized "toxics consent decrees" as court-ordered schedules for control of toxics unaddressed by EPA.[19] These amendments further extended federal involvement in state behavior in PS water policy.

Comparable to air policy, the Reagan administration made numerous attempts to revise PS water legislation, largely in the direction of greater discretion for state governments. Proposed changes that would have enhanced state authority included more emphasis on ambient goals, greater allowance of waivers of enforcement, and extension of NPDES permits.[20] These efforts were resisted by members of Congress who were protective of their power over the treatment grants program. The Reagan administration did manage to reduce the federal financial commitment to states by restricting the range of eligible POTWs and reducing the federal share from 75 percent to 55 percent in 1981.[21] Still, attempts to completely restructure the federal role were unsuccessful. The one portion of the legislation that did receive formal reauthorization during this period was the program of sewage treatment grants. Roughly $23 billion was authorized for these grants in the years 1981 through 1985. This provides sharp contrast to the 53 percent cuts in other state water grants over the same time period.[22]

Heavy federal involvement in this policy area is evident in financial expenditures. Financial involvement by the different levels of government in water policy into the mid-1980s is detailed in table 3.1. As in air policy, federal involvement declines in the Reagan years, but unlike air, the early seventies witnessed dramatic federal growth in devotion of financial resources. This corresponds to, but is exclusive of, the growth in fixed capital investment (waste treatment grants) that is evident in the last column of the table. Even aside from these massive capital amounts, federal expenditures are considerably higher than those of all fifty states combined in every year since 1972.

The 1987 reauthorization of water legislation followed considerable debate and, similar to 1972, was passed over a presidential veto.[23] The bill provides potentially greater state responsibility in several ways. One important issue is toxic pollutants. Although toxic discharges into water had been prohibited as early as the 1972 FWPCA, implementation had been slow.[24] For example, between 1972 and 1977, EPA had proposed effluent standards for only six toxics.[25] The 1987 bill mandates compliance by persistent toxic polluters by 31 March 1989 and requires states to identify those polluters and to specify enforcement strategies. The legislation also directs the EPA to identify toxics in sludge and to implement a national sludge management program. EPA has issued some interim regulations but its final standards will

Table 3.1   Water Pollution Control Funding (million 1972 dollars)

| | Federal | | | | State and Local | | | | |
	Abate & Contrl	Reg & Monitr	Res & Devel	Total	Abate & Contrl	Reg & Monitr	Res & Devel	Total	Gov Ent Capital
1972	75	79	34	188	171	66	44	281	3237
1973	124	92	59	275	155	85	31	271	3404
1974	165	118	67	350	137	96	16	249	3831
1975	214	122	60	396	147	99	15	261	4251
1976	189	114	61	364	142	128	14	284	4432
1977	191	103	61	355	117	152	13	282	4169
1978	198	125	65	388	102	139	11	252	4799
1979	194	145	73	412	115	115	12	242	4617
1980	139	189	53	381	118	107	13	238	4223
1981	101	156	59	316	117	107	6	230	3243
1982	106	135	61	302	127	100	5	232	2899
1983	175	126	67	368	134	90	4	228	2627

Source: U.S. CEQ, *Environmental Quality* (1983), pp. 616–18.

not be promulgated until at least 1991.[26] In the interim, voluntary participation by state governments is expected.

The financial structure of PS policy is also changed by the 1987 reauthorization to increase state responsibility. The legislation would provide $18 billion through 1994 for construction of waste treatment plants. The waste treatment program is supposed to convert from a system of grants to a system of loans in coming years. Specifically, the grants program is authorized through FY 1990 and then replaced over the next four fiscal years by encouraging the states to establish revolving loan fund programs to replace federal involvement. The federal funds allocated by the 1987 legislation as seed money for the conversion totals $8.4 billion. The effects of such a conversion are not yet clear. States may conceivably move to a system of privatization of waste treatment plants.[27]

Despite these recent changes, the extent of federal involvement in state PS programs exceeds that of many other policy areas. As one extensive analysis describes: "Federal policy on clean water and waste-water treatment . . . has been backed by a heavy commitment of resources, high efficacy, and . . . considerable clarity. Of the three policies we have studied closely, this one has seen the most straightforward implementation."[28] Further, a heavy federal presence in this area is likely to continue in the future. For instance, NPDES permits are valid for a maximum time period of five years. As they come up for renewal, the permittees are subject to new federal controls such as impending sludge regulations.[29]

High horizontal competition

Point source water pollution is a policy area with high potential for interstate competition. The extensive federal involvement in this policy described in the previous section is due largely to perceptions of detrimental competition between states. As in air pollution, competition between states in PS water pollution can take the form of externalities and relocations. Logically, externalities are likely since water polluters have traditionally preferred to locate by fast-moving rivers that can generate power. Those same rivers can then carry pollutants across state lines. Further, states can compete for important industries that are able to relocate and therefore able to pressure policymakers. Evidence for public recognition of this potential is apparent in the demands of environmental groups that led to the means-forcing structure described in the previous section. This section examines those demands and suggests why environmentalists were able to get the extensive federal control in PS water that they were unable to achieve in SS air pollution control. I offer this history as evidence for the fear of serious interstate competition.

The choice of technology mandates and symbolic goals instead of measurable achievement levels similar to those for air policy was deliberate and explicit. Fear of externalities and interstate competition through lax pollution controls stimulated environmental demands for national effluent limits throughout the debates on air and water pollution control. Environmental activists pushed for "national standards (which) must be prescribed for emissions from all stationary sources of pollution."[30] Criticism was even leveled on Senator Muskie, the legislative champion of environmental causes, for his critical attitude toward effluent limits.[31] Muskie, however, initially resisted these demands and stated quite explicitly his support for ambient standards rather than for effluent limits.[32] While Muskie and his supporters held the fort in stationary source air pollution legislation in 1970, two factors changed by the time of 1972 water legislation. The first of these concerns environmental interest groups and the second was the receptivity of national policymakers.

Between 1970 and 1972 environmentalists became more organized, more political, and more acceptable to other elements of society. Acceptance came from both left and right. The environmental movement in the late 1960s had been seen "as a diversion and subterfuge by civil rights, poverty, and anti-war groups."[33] By 1972 the environmental lobby consisted of a coalition of nearly thirty groups, including environmental organizations such as Friends of the Earth as well as leftist groups such as the League of Women Voters, Common Cause, the UAW, and the United Steelworkers. More conservative members of the middle class also began to support environmental causes as they realized that pollution was so pervasive that it could not be escaped simply by moving away.[34] Specifically, no location was "safe" because of the potential for externalities to cross state lines. Between May 1969 and May 1971, a period covering the original Earth Day, those ranking pollution as one of the most important domestic problems confronting the United States jumped by 24 percent to a point where few remained unsympathetic to environmental causes.[35]

Swelled by membership figures, environmental interest groups organized, shifted headquarters to Washington, and intensified political involvement. One survey by the CEQ reported that "more than half the groups had been founded during or after 1969."[36] Group activities increasingly concentrated on national political lobbying. For example, registered lobby expenditures increased between 1970 and 1971 by 57 percent for the Citizens' Committee on Natural Resources and nearly 400 percent for Friends of the Earth.[37] In the judgment of the editors of *CQ Weekly Report*, the FWPCA ranked as the second most heavily lobbied issue of 1972, whereas the CAA had been ranked twelfth in 1970.[38] Whereas 1970 testimony was supplied by the Environmental Defense Fund and the Colorado Citizens for Clean Air, the 1972 participants included the Conservation Foundation, Zero Population Growth, the Sierra Club, Environmental Action, Inc., the National Wildlife Federation, and the AFL-CIO. These are not only national organizations with considerable clout, but they testified specifically on the effluents issue.[39] This testimony was corroborated by other critical portrayals of state behavior and relocation competition contributed by analysts[40] and by the media.[41]

Support for this greater federal intervention was not unanimous. Spokesmen for many industries endorsed quality standards and opposed national effluent limits.[42] Some state and local officials also objected to increased federal control, but their position was undermined by the testimony of policymakers from other states, such as Minnesota, who warned of state susceptibility to industry relocation pressure without a strong federal presence.[43] In the end, the environmental position prevailed. When asked why their advocacy was successful in 1972 whereas it had failed two years earlier, one of the most outspoken lobbyists responded in a recent interview, "I think we were better organized in 1972."

The second factor that contributed to the greater federal control of state behavior in PS water pollution control than exists in SS air was the receptivity of the policymakers themselves to environmental warnings of high interstate competition. This receptivity in 1972 was enhanced by political pressure and

Table 3.2   Change in Water Quality (1974–1981)

	# Stations	Increase	Decrease	No Change
Ten widely used indicators				
Nitrate-nitrite as N	304	25.0%	8.2%	66.8%
Ammonia	282	11.0	10.6	78.4
Organic carbon	279	12.9	4.7	82.4
Phosphorous	301	13.0	10.0	77.0
Dissolved solids	302	22.5	16.9	60.6
Suspended sediment	289	15.2	14.2	70.6
Conductivity	305	22.6	14.1	63.3
Fecal coliform	269	7.1	12.6	80.3
Fecal streptococci	270	0.7	28.9	70.4
Dissolved oxygen	276	11.2	6.2	82.6
Average		14.4	12.6	73.0
Additional indicators				
Temperature	303	12.9	15.2	71.9
pH	304	24.5	18.4	57.2
Alkalinity	304	5.9	26.0	68.1
Sulfate	304	27.0	13.2	59.9
Calcium	304	7.6	27.3	65.1
Magnesium	304	16.4	15.1	68.4
Sodium	304	33.9	9.2	56.9
Potassium	304	22.7	13.8	63.5
Chloride	304	34.2	11.8	53.9
Silica	302	15.9	13.6	70.5
Turbidity	259	16.2	6.9	76.8
Phyloplankton	300	7.3	14.7	78.0
Average		18.7	15.5	65.7

Source: U.S. Geological Survey, *National Water Summary 1983*, p. 46.

by recognition that precedents for change in the institution of federalism itself had been established by the CAA in 1970.

The effects of political pressure are apparent in the changing nature of water pollution legislation. The Senate Public Works Committee bill of 2 February 1971 proposed state setting of effluent limits and did not represent radical departure from the CAA of the preceding year. Dissatisfaction with state control of water pollution was immediately expressed by environmentalists.[44] Such criticism was potentially devastating to the presidential aspirations of Subcommittee on Air and Water Pollution chair Muskie. Muskie had to fear losing the higher ground on these issues to the incumbent president. The Nixon administration sent out mixed signals that included apparent

Table 3.3   Active NPDES Permits as of March 1988

Facilities	Major permits issued	Minor permits issued	Total permits issued
Industrial	3,379	43,794	47,173
Municipal	3,594	11,669	15,263
Federal	145	1,151	1,296
Total	7,118	56,614	63,732

Source: U.S. GAO, *Stronger Enforcement Needed to Improve Compliance at Federal Facilities*, p. 11.

endorsement of "a base level across the country of effluent limitations" by EPA administrator Ruckelshaus.[45] Other members of the committee also had to worry about their own political ambitions. The subcommittee set about rewriting the legislation. According to Chief Staffer Leon Billings, sessions involving discussions on these issues were long and contentious.[46] However, once Muskie endorsed the concept of national effluent limits, the decision was made.

Higher federal involvement in water legislation due to the fear of interstate competition was also facilitated by the changes that had already been made to the institution of federalism. In testimony at the time and in later interviews, members of the Muskie Subcommittee showed an awareness of the dramatic breaks with federal traditions that were being legislated.[47] Billings, architect of much of both pieces of legislation, agreed that the CAA had provided the initial radical change in state pollution control prerogatives. Once the basic structure had been altered, the door was open to monumental federal control over PS pollution. In fact, anything less than a significant increase would have been viewed as stagnation. Muskie implicity admitted as such when he testified that "Without these elements (uniformity and finality), a new law would not constitute any improvement on the old."[48]

The structure created by the Muskie Subcommittee has remained basically intact since 1972. It is a structure based on fear of interstate competition through lax standards, which therefore utilizes national effluent limits on entire classes of industry. These limits are to be enforced, not changed, by the states. When PS polluters exceed effluent limits, they are deemed to be in noncompliance with the law. The EPA and the states are required to enforce permit levels before polluters are in noncompliance for two or more consecutive quarters. Polluters are required to submit periodic reports and apply for new permits every five years. This system has remained in effect since 1972 although different pollutants, particularly toxics, have been added by the 1977 amendments and the 1987 reauthorization of the act.

The Theoretical Model in the Current Context

Relative progress

Progress in the improvement of water quality has been slow. Table 3.2 displays, through reporting by monitoring stations, the overall progress in improving water quality subsequent to the 1972 legislation. The averages are revealing in that the vast majority of indicators hold constant without improvement. The largely symbolic deadlines of 1983 for fishable waters and 1985 for zero discharge passed without attainment. As of late 1987, the EPA estimated that roughly one-fourth of the nation's waterways did not meet designated uses.[49] Instances of polluted waterways still receive critical attention, as did Boston Harbor in the 1988 presidential election.

Nevertheless, progress has been made. The figures in table 3.2 do show that the deteriorating conditions prior to 1972 have been slowed or stopped in most instances. As of 1988, active NPDES permits numbered nearly 64,000.[50] These are divided between major and minor polluting facilities, based on the potential effect of discharge on relevant bodies of water. Table 3.3 shows the division of these permits between industrial, municipal, and federal facilities. Severely polluted bodies of water such as Lake Michigan and the Chesapeake Bay have been significantly improved. In many cases, slow progress is due more to neglected nonpoint sources of runoff rather than to the regulated point sources. Most national surveys suggest that enough progress has been made in controlling point sources that overall water quality has been basically maintained in the face of a growing population.[51]

A couple of examples show that the pace of progress has been increasing since 1980. First, whereas roughly 10 percent of industrial facilities were in noncompliance in 1979, that figure is recently down to only 6 percent.[52] Only in the area of federal facilities has improvement been nonexistent.[53] Second, at the end of the 1970s, only half of relevant municipalities enjoyed secondary water treatment and 87 percent of the expensive wastewater treatment plants were cited as violating their permits.[54] By 1986 these problems had been corrected to the extent that 127 million Americans were served by secondary treatment, an increase of nearly 50 percent over 1972.[55]

Hypothetical impacts of state governments

The theoretical model developed in chapter 1 generates several hypotheses concerning the impact of state governments on PS water policy failures and possibilities. The first hypothesis suggests that some matching between severity and response is evident and that variance in state behavior is lower than that in ss air policy. The second hypothesis predicts high potential for horizontal competition can make even leading state programs reluctant to exceed federal standards and guidelines when they exist. The third hypothesis argues that high vertical involvement facilitates the communication and coordination of leading program efforts across state lines. Evidence for these hypotheses is presented with data on overall state behavior and an in-depth analysis of one leading state program: North Carolina.

Overall State Behavior in Point Source Water Policy

Behavior of the states in this policy area is crucial to progress in the control of ps water pollution. Evidence for the behavior of all fifty states is presented in terms of delegated authority, expenditures, and enforcement.

Delegated authority

One way to measure state programs in ps water pollution control is to assess the extent of authority delegated to the state level by the federal government. When the EPA deems a state worthy of delegation, the state program is given responsibility to issue permits, monitor compliance, and enforce effluent limits on federal facilities.[56] The positive side of this operationalization is that it represents a "fundamental measure of state commitment to surface water protection."[57] The negative side of relying too extensively on such a measure is that the delegation of federal authority could conceivably be determined by aspects other than merit. Nevertheless, tests on this variable show some evidence for state responsiveness to the presence of polluters.

Table 3.4 uses a dichotomous dependent variable. The value of 1 is given if a state program has received federal approval for its NPDES program, its regulation of federal facilities, and its pretreatment program.[58] This dependent variable is tested in a simple logit equation against measures of the independent variables described in chapter 1. Briefly, *fish* is a rough measure of water quality indicating the percentage of fishable waters in the state at the end of the 1970s. The variable *mugsp* reflects the presence of potentially heavy polluting segments of industry in the state's economy. Relative affluence of the state is measured by personal income per capita (*pipc*). The political culture of the state is represented by *piindex*. Finally, because of the massive amounts of grant money involved, federal financial presence is measured in two ways. The measure including this potentially deterministic amount is indicated by *fwtgpc*, which stands for federal waste treatment grants per capita through the early 1980s. Federal involvement is also measured by federal funds allocated to states under Section 106 (*f106pc*) for purposes exclusive of waste treatment grants.

Table 3.4 Delegated Stationary Source Water Authority (dependent var. = complete water program authority)

Independent Variables	Logit (w.t.g.)	Logit (106)
constant	4.91	3.39
	(7.76)	(6.88)
fish[a]	−0.78	−0.74
	(0.77)	(0.74)
mugsp[b]	0.16*	0.17*
	(0.06)	(0.06)
pipc[c]	−0.16	−0.08
	(0.22)	(0.18)
piindex[d]	−7.66	−7.58
	(5.74)	(5.73)
fwtgpc[e]	4.06	—
	(7.67)	—
f106pc[f]	—	1.20
	—	(2.89)
Chi square	17.72	17.60
sig level	0.005	0.005
*significant at the 0.05 level		

a. % of state waters fishable or swimmable
b. manufacturing/utilities as % of gross state product
c. personal income per capita
d. index of party competition and turnout
e. federal waste treatment grants per capita
f. federal 106 grants per capita

The results in table 3.4 indicate some correlation between severity of the problem and policy response. The variable representing polluting presence is significant in both equations, suggesting that the states with high PS pollution potential are also the states with federally accredited programs. The variable for water quality is not significant, but similar to comparable measures in air policy, this may be more of a methodological problem than a meaningful result. Specifically, water quality is affected by point as well as nonpoint sources.[59] The equations are statistically significant above the 0.005 level.

Expenditures

Evidence for state responsiveness in PS water expenditures is less convincing. Table 3.5 uses state expenditures in FY1986, the latest year available, as the dependent variable. The independent variables are the same as those listed

Table 3.5   State Stationary Source Water Pollution Control Expenditures
(dependent var. = water expenditures per capita FY 86)

Independent Variables	OLS (w fed)	OLS (w/o fed)
constant	0.65	−0.92
	(1.86)	(1.84)
fish[a]	−0.12	−0.24
	(0.18)	(1.89)
mugsp[b]	0.31	0.97
	(1.34)	(1.38)
pipc[c]	0.91	0.97*
	(5.72)	(0.47)
piindex[d]	0.25	0.37
	(0.81)	(0.86)
fwtgpc[e]	0.48*	—
	(0.20)	—
R square	0.21	0.10
*significant at the 0.05 level		

a. % of state waters fishable or swimmable
b. manufacturing/utilities as % of gross state product
c. personal income per capita
d. index of party competition and turnout
e. federal waste treatment grants per capita

above. The table shows two equations so that the influence of the waste
treatment grants program can be isolated. The grants variable and the measure of affluence are in fact correlated. Indeed, the wtg variable dominates the
first equation, but when it is omitted, the variable measuring relative affluence (pipc) absorbs its significance. The R-square terms are not very impressive for either equation. This could simply imply that these independent
variables do not explain spending decisions. This result may also reflect the
low variance in state expenditures on PS water policy which was shown in
table 2.6. Indeed, Hypothesis 1 predicts that variance between state behavior
in this policy area is much lower than that in SS air policy.

The waste treatment grant program itself may skew state decisions away
from water quality needs. The technicalities of this pork barrel, funding for
construction costs but not for operation and maintenance, are reflective of a
congressional response that is more concerned with electoral consequences
than with operational effectiveness. Members of Congress can take credit for
construction of waste treatment plants and wash their hands of the ensuing
operation. On the other hand, because these grants subsidize capital but not

Table 3.6   State Reception of Waste Treatment Grants (dependent var. = federal waste treatment grants per capita)

Independent Variables	OLS estimation
constant	−2.98*
	(1.36)
fish[a]	0.19
	(0.14)
mugsp[b]	0.12
	(0.10)
pipc[c]	1.83*
	(0.34)
piindex[d]	1.02
	(6.36)
epaa[e]	−0.21
	(0.21)
R square	0.45
*significant at the 0.05 level	

a. % of state waters fishable or swimmable
b. manufacturing/utilities as % of gross state product
c. personal income per capita
d. index of party competition and turnout
e. EPA actions for noncompliant state behavior

operating costs, state officials have incentives to build expensive, capital-intensive plants but little incentive to concern themselves with subsequent operations.[60]

One important question suggested by the significance of the waste treatment grant program in state spending is whether or not states with a heavy polluting presence are those that are willing to spend the necessary amounts in matching funds to obtain these grants. Table 3.6 shows water treatment grants per capita as determined by the independent variables already described, with one change. Since federal influence obviously cannot be the same variable on both sides of the equation, the federal presence is measured here in EPA actions against the states for noncompliant behavior (*epaa*). Again, relative affluence (pipc) is deterministic. Water quality (fish) and polluting presence (mugsp) are not significant but nevertheless important variables in the equation. The t-statistics for these variables do not reflect significance, but neither do they suggest that pollution has no bearing on whether or not states obtain portions of this massive grant program. Since the universal aspect of the grant program legislation mandated that each state would receive at least one-half of 1 percent of the funds, perhaps a stronger showing for these variables should not be expected.

Enforcement

The state record in enforcement against PS water polluters is also mixed. As in air, enforcement by state authorities in water pollution traditionally relied largely on self-monitoring by polluters and the occasional Notice of Violation intended mainly to bring the offender back in line, not to punish past behavior or prevent future behavior.[61] This appears to be changing as some states apply stiff fines and penalties. North Carolina, as described below, does not hesitate to punish noncompliant behavior. Following passage of a 1986 law allowing heavy fines, Massachusetts assessed nearly $700,000 worth of penalties in the first six months.[62] The major exception to increasingly stringent enforcement by state governments is predictable considering the heavy federal presence in PS control described earlier. States are still reluctant to enforce limits against federal facilities. In forty-six cases of federal facility violations over fiscal years 1986 and 1987, only eight received timely action from the proper authorities.[63]

Table 3.7 displays a more systematic assessment of state enforcement efforts. The dependent variable utilizes data compiled by the FREE organization on state water programs that includes monitoring, compliance, and backlogs. The variable is actually a combination of FREE's rankings of surface water and ground water programs. This is done because both are impacted by PS polluters. The independent variables are as described above, retaining the epaa variable because of multicollinearity between affluence and waste treatment grants. Two columns are presented, one for an OLS estimation of the equation and the second column for a logit analysis with the dependent variable divided into strong and weak programs. Both equations are fairly strong. The higher variance in the dependent variable in the OLS equation allows a strong showing by the federal variable. This variable's negative sign reflects the imposition of federal actions on weaker state programs. Both equations show the importance of personal income per capita in developing strong programs. The significance of this variable is consistent with the argument that relatively affluent states may be more willing to risk losing PS polluters through relocation than are poorer governments. The strong showing of the income variable in all of these results suggests a caveat about the resurgence of state governments. Responsive behavior is more likely when it is affordable.

Finally, mention should be made of perhaps the most important area of PS policy enforcement by state governments in coming years. Enforcement in

Table 3.7 Rankings of State Stationary Source Water Programs (dependent var. = surface + ground water rankings)

Independent Variables	OLS	Logit
constant	0.64	−9.21
	(7.65)	(7.64)
fish[a]	0.51	0.23
	(0.78)	(0.73)
mugsp[b]	0.37	0.46
	(0.57)	(0.55)
pipc[c]	0.47*	0.55*
	(0.19)	(0.22)
piindex[d]	−0.33	−0.95
	(0.36)	(0.58)
epaa[e]	−0.58*	−0.39
	(0.12)	(0.24)
R square	0.44	
Chi square		21.08
sig level		0.005
*significant at the 0.05 level		

a. % of state waters fishable or swimmable
b. manufacturing/utilities as % of gross state product
c. personal income per capita
d. index of party competition and turnout
e. EPA actions for noncompliant state behavior

the area of toxic substances shows that states are beginning to make some progress. Toxicants are pollutants that can seriously damage biological systems in living organisms, to the point of causing death. Similar to SS air pollution control policy, this issue is left largely up to state governments. As of 1986, all states had narrative requirements for toxic pollution, but more importantly, more than thirty states had established numeric criteria for toxics.[64] In addition, thirty-seven states require at least some industries to conduct bioassays, testing of toxic effects on organisms in the water.[65]

Overall performance of the states

Some states have developed professional programs to control PS water pollution. The slow pace of progress toward attainment of water quality goals is more attributable to nonpoint than point sources of pollution. Still, PS control is more likely if states have resources available to afford strong programs. The following section examines one leading state in this particular policy area.

Leading State Behavior in Point Source
Pollution Control

Leading state behavior in PS water pollution control is significantly affected by high potential interstate competition and heavy federal involvement. I describe the fundamentals of one leading state program and then consider some innovative developments. Data used in the preceding analysis allow identification of North Carolina as one of the leading states in PS water pollution control. North Carolina has had the complete program authority described for table 3.4 since 1984.[66] In the rankings used in table 3.7, the state's surface water program scored 10 out of 10, with no other state receiving even a 9.[67] The groundwater program received an 8 out of 10.[68]

Description

Water, and the control of water quality, is abundant in the state of North Carolina. Over 300 miles of ocean coastline and 3,200 square miles of estuaries are complemented inland by 37,000 miles of freshwater streams and rivers, 305,000 acres of lakes, and countless acres of wetlands.[69] The state has implemented practices to control the quality of these bodies of water since 1972. Responsibility for state efforts is housed in the Division of Environmental Management (DEM) within the Department of Natural Resources and Community Development. This section describes the fundamental components of the North Carolina PS program.

Central to the PS water program is the system of NPDES permits. As of 1988, the DEM monitored 3,584 point source permits within the state. In terms of the categories cited in table 3.3, 305 (120 major) are municipal, 47 are federal, and the other 2,301 are industrial and other dischargers (94 major).[70] The DEM requires permits for every type of waste disposal, from single residences to factories, that might affect water quality. One leading administrator asserted that coverage was much more thorough than in other states, an assertion supported by other sources.[71] Table 3.8 shows some of the fees that accompany the permits and help fund the water quality program.[72] Several points are apparent from the table. First, with over three thousand permits, the fee schedule generates considerable revenue. Annual revenues from fees are expected to exceed $2 million, or roughly 25 percent of the water budget.[73] Second, the fee framework is comprehensive enough to allow thorough coverage. Third, effort is rewarded as each polluter re-

Table 3.8   North Carolina Stationary Source Water Permit Fees

Category	Application		Annual	
	Standard	Renewal	Standard	Compliance
> 10 mill GPD (gallons per day)				
industrial	$400	200	1200	900
domestic	400	200	900	675
1 to 10 mill GPD				
industrial	300	150	800	600
domestic	300	150	600	450
100,000 to 1 mill GPD				
industrial	250	125	400	300
domestic	250	125	300	225
< 100,000 GPD				
industrial	200	100	300	225
domestic	200	100	225	150
single family dwelling	120	60	—	—
most general permits	200	0	200	0

Source: N.C. *Administrative Code*, Sec. 15A NCAC 2H.0100, p. 5.

ceives a 25 percent discount on the annual administration and monitoring fee if its facility is in compliance.

In addition to comprehensiveness, the North Carolina permit process is stringent. The state heavily utilizes the ambient option described in a preceding section. The DEM explicitly states that "The primary control on point source pollution in North Carolina is the imposition of water quality-based effluent limits in NPDES permits."[74] As detailed in the state code, the state uses federal guidelines for effluent limited streams. However, when those guidelines do not allow attainment of designated water quality, then the DEM calculates its own limits.[75] Two different high-ranking officials each estimated that 75 percent or more of the state's waterways are subject to water quality standards that impose more stringent limits on polluters than optional federal guidelines do. This figure is much higher than in most states. The North Carolina standards are based on increasingly sophisticated mathematical modeling and a hearing process for the determination of designated uses for bodies of water.[76] The state utilizes some 340 ambient monitoring stations to collect data for their models and to detect quality trends in conducting nearly one thousand wasteload allocations per year.[77]

Besides comprehensiveness and stringency, the North Carolina NPDES process is aggressive. Administrators used this term frequently to emphasize the sincerity of their efforts. The DEM utilizes computer targeting, inspections, and auditing to determine if and when enforcement actions are necessary. Enforcement actions include fines, assessments, and even revocations of permits. As one chief enforcer asserted, the state will "hit 'em (violators) hard if they need to be hit hard." Results have been significant. According to one outside authority, only 1 percent of major nonmunicipal dischargers were in significant noncompliance in late 1986.[78] Overall compliance of major dischargers was roughly 94 percent in 1985 and has been significantly improved in subsequent years.[79] Not only are compliance rates among the highest in the nation, but the state is a leader in actually achieving zero discharge. The DEM is increasingly requiring polluters to show that no nondischarge alternatives are economically feasible before issuing permits.

Overall water quality results reflect the effectiveness of the PS permit program. More than 90 percent of the state's streams and rivers were assessed for the 1988 305B progress report. Nearly 68 percent of evaluated streams fully support and 28 percent partially support their designated uses. Less than 1 percent of lakes, reservoirs, estuaries, and sounds do not support designated uses.[80] Most of this progress has occurred in the last two decades. During that time, for example, more than two hundred municipalities have received federal or state funds to improve water supply, so that more than 80 percent of the state's population now has more than secondary treatment.[81] Compliance rates on industrial facilities are high and getting higher. According to the 305B evaluation, nonpoint sources account for more than 90 percent of the impairment that still exists.[82] Point sources of water pollution are, for the most part, well under control.[83]

In addition to the permit program, North Carolina is showing progress in other water policies. The DEM recognizes that nonpoint sources of pollution have not received as much attention as point sources have in the past. In fact, these priorities are attributed (as described in chapter 5) to federal legislation.[84] However, the state has made significant progress in its NPS efforts in recent years with agricultural cost-sharing and stringent coastal stormwater rules that federal authorities have so far been only talking about.[85] The state is also actively pursuing the major cause of groundwater problems with a $1 million inventory and assessment of leaking underground storage tanks.[86] Positive results from these efforts are crucial to continued improvement in state water quality.

Leadership

North Carolina policymakers have shown leadership in several aspects of PS water pollution control. These include organizational decentralization, basinwide assessments in permits, toxicity regulations, and pollution prevention. Most of these innovations have been developed in the last five to seven years, a time period consistent with the hypothesized resurgence of state governments. These innovations fill gaps in the existing federal structure and are being rapidly disseminated to other states and even to the federal level.

The North Carolina DEM relies heavily on its regional offices for implementation and enforcement. North Carolina has seven regional offices, each organized similarly to the state office with branches for groundwater, water, and air control.[87] The regional offices are responsible for inspections, issuance of notices of violations, and enforcement of penalties. Officials within these offices are the public employees who are in actual contact with PS polluters. To ensure that "capture" does not occur at this local level, the regional officials answer to two different bosses, the regional supervisor (for all media) and the state water quality administrator. Open lines of communication are emphasized both across regions and between the regions and the central office. This decentralization facilitates thoroughness and responsiveness to local circumstances. One high-ranking state enforcer, a former regional operative and supervisor himself, was proud to note that most states are more centralized and therefore less effective. Even so, many states are currently developing greater reliance on decentralized offices precisely because decentralization does allow more specific responses to problems.

Another area of innovation in North Carolina's PS water pollution control program is basin permitting. As of the start of 1990, the state is converting to an arrangement in some instances whereby all discharger permits in the same basin will expire at the same time. This allows across-the-board cutbacks in permitted discharge if necessitated by water quality in the basin. Again, this conversion becomes possible because of greater sophistication in water quality modeling.

DEM administrators are quite proud of their aquatic toxicology program. North Carolina was among the first, if not *the* first, states to apply toxic water permits in a regulatory capacity. The DEM requires certain NPDES permittees to conduct tests to determine the effect of effluents on specified organisms.[88] The actual test procedures are specified within each NPDES permit or administrative letter. Test results are recorded as pass or fail.[89] The DEM itself conducts on-site toxic evaluations of ambient water quality as well as of effluents.[90] The redress of toxic pollutants has been mandated by federal law for years.[91] However, it has only been a reality since North Carolina in 1987 and other states developed the capability to test toxic effluents. North Carolina has set numeric criteria for fifteen nonorganic pollutants and sixteen organic pollutants, more than in most other states.[92] Further, the state has an extensive bioassay program.[93] According to Ken

Eagleson, one of the architects of the North Carolina effort, toxic regulation stimulated considerable opposition by municipalities and at least implicit threats of relocation by industries but was heartily applauded by federal authorities. Dissemination of this innovative response is apparent in the efforts of other states, especially those neighboring North Carolina. For example, South Carolina also adopted pass/fail toxic tests in 1988.[94]

Finally, let us return to the story of the Pollution Prevention Program that began this chapter. Like any innovation, the PPP started as an idea. In a 1982 North Carolina symposium, local academics proposed public efforts to encourage industrial restructuring toward cleaner technology and more pollution abatement.[95] The state appropriated $300,000 to research pollution prevention in 1984. The PPP was adopted by the Department of Natural Resources shortly thereafter and has since become a significant part of the state's pollution control efforts. The PPP provides technical assistance in the form of information and on-site (roughly one/week) advice to polluters seeking help, financial assistance in matching grants for pollution prevention ($880,000 since 1985), and research and education projects.[96] Gary Hunt has been with the PPP since the beginning and is now the director. In a recent interview, he described why the PPP receives so much support. The program has a good relationship with regulators because they provide the "best referrals." In other words, when a polluter is in trouble with the DEM, they often come to PPP for advice. Polluters support PPP because Hunt and his employees provide free support, which can indeed save money and deter regulatory penalties. Similar programs have been developed in other states and are now being pushed at the federal level for a simple reason.[97] As Hunt says, "The bottom line is economics, it's gonna save money." North Carolina is not only leading the states, but Hunt's former boss was recruited by EPA to coordinate federal efforts.[98]

Two aspects of these innovations should be emphasized. First, innovative state actions are occurring in PS water policy in areas where the federal government expects results but has not become involved, such as toxics and prevention. State leadership does not exceed federal guidelines unless specifically allowed, as in ambient monitoring. In fact, North Carolina even has explicit legislation (the Hardison Amendment) that prohibits state policymakers from exceeding federal authority in these areas. As one official told me, "When they (the federal government) don't specify something, that's when we can jump in." Some of these innovations have spread rapidly. Most states are now tackling the toxic problem and more than thirty-five states have pollution prevention programs.[99] Further, the federal government is also adopting some of these innovations.

The second conclusion regarding these innovations is less positive. The reluctance to exceed federal guidelines in areas where the federal government has become involved but has not defined the state's role, such as nonpoint source (NPS) control and sludge disposal, has slowed progress toward quality goals. Slow development of NPS programs has prevented attainment of water quality goals in many states, including, as mentioned earlier, active states such as North Carolina. This policy area is considered more fully in chapter 5. While it can be argued that NPS problems are distinct from PS programs, the same cannot be said for sludge disposal. As POTWs continue to increase the annual generation of sludge, state participation in interim federal programs remains "low." As of late 1989, only eight states were willingly participating fully in disposal efforts and their approaches displayed considerable variance.[100] As federal guidelines remain vague, interim, and voluntary, state policymakers may find it difficult to muster the resources and motivation necessary to fill in this gap in water policy.

Explanation

Why has North Carolina been a leading state in PS water pollution control? North Carolina has long been considered more progressive in many areas than most of its southern neighbors.[101] The state populace is quite literate and environmentally conscious. Academics provide a significant presence with so many universities and researchers (Research Triangle Park) in close proximity to the state capital. Several of the policymakers I interviewed cited the availability of expertise and the presence of academics on advisory boards as contributing to progressive ideas. Eagleson attributed the development of the water toxics program to a fairly unique combination of expert technicians and dedicated administrators. Finally, state political institutions have been, for the most part, supportive. For example, the 1989 state assembly passed numerous pieces of environmental legislation with direct impact on DEM efforts, including watershed protection, stormwater runoff rules, and offshore oil requirements.

Conclusions

The dimensions of federalism governing point source water pollution control policy can be characterized as high in vertical involvement and high in horizontal competition. Unlike SS air policy, the federal government has been willing to specify means of attainment as well as overall quality goals in this policy area. Similar to SS air policy, polluters can viably threaten relocation as important single entities. Further, since many point source polluters have traditionally depended on currents of waterways to generate power, their pollutants may well be deposited in bodies of water that cross state lines.

The effects of these dimensions on state leadership are apparent in evidence supporting the three hypotheses of chapter 1.

Hypothesis 1. Some correlation exists between severity of the problem and responsiveness at the state level in PS water policy. Matching is apparent in the delegation of responsibility by the federal government to qualified states (table 3.4). Overall matching of response to need is significantly influenced by the availability of resources (table 3.7). Matching behavior is more likely when states have enough resources to be less concerned about losing important constituents to relocation.

Hypothesis 2. The high potential for interstate interaction affects not only the importance of available resources in Hypothesis 1 but also the willingness of leading state programs to innovate. State leaders display creativity in filling gaps in the federal structure but reluctance in exceeding federal guidelines where they exist. North Carolina policymakers have developed coastal stormwater rules, toxic regulations, basin permitting, and decentralized application of ambient standards as innovations that do not directly exceed federal guidelines. On the other hand, states have been slow to regulate sludge disposal and polluting federal facilities.

Hypothesis 3. High vertical involvement facilitates dissemination of innovations. Communication and coordination of creative programs has been impressive. North Carolina's efforts in basin permitting, aquatic toxicology, and pollution prevention have been rapidly adopted by other states and even by the federal government.

State governments have made serious progress in the control of point source water pollution since 1972. Much of the innovative behavior in this policy area is taking place at the state level. Some of the state policymakers I interviewed felt that the federal government was lagging behind in areas such as pollution prevention. Communication and coordination of these innovations could be enhanced by regional seminars, voluntary transfers, and on-site workshops. Such efforts would not only keep the federal government informed but may help to lessen the gap between affluent and less-affluent states.

4   Mobile Source Air Pollution Control

From the highest peak in the continental United States, you can see Los Angeles. Mount Whitney is located in the Sierra Nevada over one hundred miles northeast of L.A. Looking southwest from the summit you see an ugly brown cloud that spoils an otherwise beautiful horizon. That cloud represents the worst manifestation of all the activity in Tinsel Town, smog from thousands of motor vehicles. When I asked Deputy Executive Officer Tom Cackette, of the California Air Resources Board (CARB), why his state traditionally led the rest of the nation in automobile pollution controls, he responded immediately with "L.A."

This chapter describes the leadership of the states in the control of mobile source air pollution. In this policy area, the dimensions of federalism stimulate an efficient match between problem severity and state response. Further, some states, particularly California, have developed comprehensive, independent responses to tragic situations such as the smog in Los Angeles. These responses explicitly exceed federal standards and are rapidly emulated by other leading states.

The Role of the States in Mobile Source
Air Pollution Control

Mobile source (MS) air pollution is a serious problem. Gasoline-powered vehicles emit hydrocarbons (HC), carbon monoxide (CO), nitrogen oxides (NOX), and particulates. The amounts of pollutants depend on the size, design, and condition of the engine as well as the load being pulled. Automobiles contribute roughly two-thirds of all vehicle pollutant emissions.[1] Smog is principally composed of ozone, which is formed by a chemical reaction of volatile organic compounds (VOC, hydrocarbons and nitrogen oxides) in sunlight. About 31 percent of VOC emissions in 1987 were contributed by transportation sources.[2] Effects on humans from MS emissions include reduced visibility, impaired breathing, and potential damage to heart functions. Even though an explicit federal policy structure to control vehicle emissions has existed since 1970, cities like Los Angeles remain plagued by MS pollution.

Mobile source emissions are obviously a public goods problem. Each vehicle contributes pollutants to air that all have to breathe. Material incentives exist for each driver to avoid costs on his or her own vehicle while hoping that others will reduce emissions from theirs. The usual American solution to a public goods problem is government intervention. In the case of MS pollution control, government intervention has involved extensive federal involvement. The already low relative potential for interstate competition is thus further reduced.

High vertical involvement

Federal policy involvement in this area was stimulated by initial government action that originated in the states, California in particular. As early as the 1950s, California officials asked the automobile industry to do something about the smog problem in Los Angeles. The lack of response by manufacturers led to state actions in the early 1960s that progressed from voluntary installation of crankcase devices in 1961 to required exhaust controls for 1966 cars sold in California. These actions prompted 1965 federal amendments to existing air legislation that directed the U.S. Department of Health, Education, and Welfare to establish emission standards for motor vehicles. In 1968 HEW set standards for HC and CO for cars and light trucks. This set the stage for formal institutionalization of the process two years later.

The 1970 Clean Air Act provides the basis for extensive federal control of MS emissions. Unlike most of the provisions in the act for stationary source air polluters, the legislation specifies required percentage reductions in HC, CO, and NOx emissions from mobile sources. This differentiation between mobile and stationary sources is interesting considering the relatively equal contribution of these two categories of air pollution sources. Authorities at the time recognized that moving sources were responsible for about 42 percent of total emissions of the five major pollutants.[3] The legislation called for 90 percent reductions in HC and CO by 1975 from 1970 levels and 90 percent reductions in NOx by 1976 from 1971 levels. Cars were to be built that met these standards for five years or 50,000 miles, whichever came first. All parties concerned recognized these standards as "technology-forcing" in that they were beyond existing manufacturing capabilities.[4]

Several reasons explain why the CAA was so explicit on MS pollution. First, consensus existed that automobiles were the largest source of cited emissions and responsible for the ugly haze in Los Angeles and other places.[5] Second, as the California experience showed, auto executives had displayed "little willingness to respond voluntarily" and thus stimulated explicit directions.[6] Third, the auto lobby did not fight the national intervention as much as might have been expected largely because they feared fifty diverse sets of state standards even more than one national set.

These significant federal controls were further strengthened by the 1977 CAA amendments. By then, experts blamed automobiles for 40 percent of HC, 84 percent of CO, and 39 percent of NOX emissions.[7] The amendments set stringent standards for new automobiles, as shown in table 4.1. The legislation allowed only the state of California to set independent tailpipe standards. However, the other states were allowed to adopt California's standards once established. Obviously, the required percentage of reductions shown in table 4.1 were immense. Amendments also tightened standards for emissions from heavy-duty engines and for particulates from diesel engines. Policymakers required gas stations to make available unleaded gas to be used in catalytic converter-equipped vehicles. In addition, the law recognized the effect of emissions at higher altitudes and allowed EPA to set higher standards for cars sold in such locales (Sec. 202-f). The 1977 CAA further required states to revise their State Implementation Plans (SIPs) by 1 January 1979 showing how attainment of ambient standards would be achieved by 31 December 1982 (with limited exceptions). Finally, provisions were made for mandatory usage of inspection and maintenance (I/M) programs.

I/M programs thus became an integral part of MS pollution control in 1977. Data showed that 80 percent of the 100 million automobiles in the country did not meet existing standards, a total attributed largely to maladjustment (47 percent) and tampering (18 percent).[8] One proposed solution for inadequate in-use performance was for states to set up mandatory I/M programs involving emissions tests to detect and encourage repairs of high-polluting vehicles. The proposal was not entirely new. The 1970 CAA had allowed I/M programs. In addition, a 1973 decision by the U.S. Court of Appeals in Washington endorsed such procedures in states adversely affected by MS pollution.[9] However, independent state adoption of I/M programs prior to 1977 was not widespread due largely to local public disapproval.[10] Some authorities endorsed federally imposed mandatory I/M programs as effective enforcement of in-use emissions controls.[11]

The federal response to these demands is fairly explicit. While setting attainment deadlines for ozone and carbon monoxide for the end of 1982, the law allowed for five-year extensions (until 31 December 1987) for ozone

Table 4.1   1977 Light-Duty Vehicle Emission Limits (grams/mile)

Pollutant	Without Control	1977 Standard	Percent Reduction
Hydrocarbons	8.8	0.41	95%
Carbon Monoxide	87.0	3.40	96%
Nitrogen Oxide	3.6	1.00	72%

Source: U.S. EPA, *Progress in the Prevention and Control of Air Pollution in 1986*, p. IX-1.

and hydrocarbon nonattainment areas if revised SIPs included a specific schedule for implementation of an I/M program. The definition of an acceptable I/M plan, according to EPA policy of July 1978, is one that would achieve 25 percent reductions in HC and CO emissions by 31 December 1987 compared to what would have been emitted without the I/M program. This goal was based on the accomplishments of the existing New Jersey I/M program.[12] EPA enforcement of this provision has been inconsistent in terms of its stringency and application.[13] Still, EPA has set up an I/M audit program. In 1986, for example, the federal agency conducted thirteen audits and six follow-ups to pinpoint deficiencies in existing I/M programs.[14] Other details of I/M programs, such as time intervals between inspections and variance on standards by age of the car, were left to state and local policymakers.[15]

In addition to emission controls and mandated I/M programs, the federal government has had an impact on automobile pollution with the imposition of the Corporate Average Fuel Economy (CAFE) requirement. Higher gas mileage means less burning of fossil fuels. The CAFE requirement was set by Congress at 27.5 mpg as an average for all new cars produced by a manufacturer to be achieved by 1985. One provision allows the Department of Transportation to amend the requirement if circumstances change. Indeed, in response to the demands of auto manufacturers, the standard for 1986–88 was lowered to 26 mpg and for 1989 to 26.5. Auto producers criticize the CAFE for reducing competitiveness and for targeting "the source of a remarkably small share of these (greenhouse) gases on a worldwide basis" even while admitting that American vehicles contribute 5 percent of the world's CO_2 emissions.[16]

The federal government has continued to maintain a strong role in the control of MS emissions. Efforts prior to the comprehensive 1990 Bush administration proposals have occurred in four major areas. First, the EPA promulgated a rule taking effect 1 January 1986 reducing the lead allowed in gasoline from 1.1 gram/gallon to 0.1 gram/gallon.[17] Second, the agency proposed cuts in commercial gasoline volatility to control evaporative emissions. Hydrocarbons released by gasoline vapors are especially problematic during the summer heat. In 1987 EPA proposed Phase I reductions of summer gasoline volatility to 10.5 pounds per square inch by 1989 and Phase II reductions to 9 psi by 1992.[18] Third, EPA has submitted a draft proposal requiring additional onboard controls to reduce hydrocarbons released in refueling.[19] Finally, EPA proposed to require ozone nonattainment areas to reduce the emissions of hydrocarbons annually.[20]

The 1990 Clean Air Act maintains a high level of federal involvement in this policy area. The legislation was originally proposed by the Bush administration in 1989.[21] The bill went through numerous changes in the Congress before becoming law on 15 November.[22] The 1990 Clean Air Act lowers emission limits in hydrocarbons and nitrogen oxides by substantial amounts, proposes pilot programs for alternative-fuel cars, and requires cleaner-burning gasoline to be available in the smoggiest cities. The provisions of this bill concerning stationary sources of air pollution are discussed in chapter 2.

The cost effects of these federal requirements should not be underestimated. As early as 1981, new cars averaged $582 worth of emission control equipment.[23] The preproduction certification program for new cars is quite extensive, involving testing of prototypes, EPA review of test results, and identification and correction of unacceptable emission levels.[24] The costs to states for inspection programs run in the millions of dollars. For example, Wisconsin's program is estimated to total $9.2 million annually.[25]

Low horizontal competition

Horizontal competition between states is relatively low in this policy area. Mobile source pollution involves relatively little impact from interstate transfer of polluters or pollutants.

Automobile drivers enjoy little leverage through threats of relocation. Even if emission limits are stringent, vehicle operators are too numerous and too unorganized to threaten relocation by suggesting a massive transfer of drivers from one state to another. Unlike the loss of large industries or utilities, the loss of one driver to another state will be unnoticed. National organizations such as AAA could organize actions within individual states, but such behavior has been relatively ineffective.

Interstate externality effects from MS auto pollution are also relatively low compared to that from stationary sources. A state cannot encourage all of its drivers to mass on the downwind boundary of the state so that emissions are externalized to neighboring locations. Much of the ozone that does occur is trapped by geography and temperature inversions in the cities where it originates. Of the heavy acid rain in northeastern states, 70 percent is estimated to be caused by sulfur-based, nonautomotive sources and the remaining 30 percent is nitrogen-based emissions that are split between utilities, industries, and transportation sources. One estimate attributes as little as 5 percent of acid rain deposition to automobiles.[26]

The Theoretical Model in the Current Context

Relative progress

Most authorities agree that substantial progress in the control of mobile emissions has been accomplished but also that progress has not come cheaply. Since 1977 ozone emissions have been reduced by 21 percent, carbon monoxide by 32 percent, nitrogen dioxide by 14 percent, and airborne lead by 87 percent.[27] In return, Americans pay more for vehicles, more for fuel, and more for services than they have in the past.[28]

Progress slowed during the 1980s. The Reagan administration relaxed regulations such as the CAFE requirement.[29] The summer of 1988 made millions of Americans realize that Los Angeles is not the only city to suffer from smog. Particularly high temperatures contributed to serious urban pollution problems. In the nation's capital, for example, the levels of ozone peaked at 50 percent higher than the federal standard, one of the highest levels in decades. Mobile sources produced the vast majority of the offending pollution.[30] In all, nearly one hundred urban areas occasionally suffer from ozone levels in excess of standards set to protect the public health.

Hypothetical impacts of state governments

The theoretical model developed in chapter 1 generates several hypotheses concerning the impact of state governments on MS air policy failures and possibilities. The first hypothesis suggests that matching between need and response is quite high. Heavy federal influence and low interstate interaction ensure that the severity of the problem is a significant determinant of variation in state behavior. The second hypothesis argues that low horizontal competition enables leading state policymakers to exceed federal guidelines. The third hypothesis predicts that the efforts of leading state programs can be quickly emulated.

Overall State Behavior in Mobile Source
Air Pollution Control

State governments have less discretion in controlling MS air pollution than they do for most stationary sources.[31] National manufacturers are responsible for mandated emission reductions. Nevertheless, the CAA does contain

Table 4.2   Early I/M Auto Pollution Programs

Locale	Date	Description
Arizona	1974	Network of 12 I/M stations. Mandatory $5 tests starting in 1976 but repairs optional until 1977. 1 million cars inspected with 50% rejection rate in first year. Public outcry lowers later standards.
California	1973	Statute calls for local programs. Voluntary program in Riverside between 9/75 and 2/76 ran rejection rate of 34%. Competitive bidding for L.A. program awarded on 6/30/77 but unfinished.
Chicago	1973	In response to EPA request, city began program which requires inspection but provides no powers of enforcement.
Cincinnati	1973	In response to EPA request, city began program which is largely ineffectual since Kentucky cars (80,000 in the city each day) are exempt.
New Jersey	1974	Mandated by state law. Standards will vary by model year with rejection rates ultimately to reach 30%.
Oregon	1973	Voluntary program with widespread acceptance encouraged by bumper stickers. Mandatory tests started in 1975 with certificate costs of $5 but no charge for failed tests or retests.
Rhode Island	1976	Legislature passed bill calling for private contractors but public outcry put plan on temporary hold.

Source: U.S. EPA, *A Review of Vehicle Inspection and Maintenance Programs in the United States;* U.S. GAO, *Better Enforcement of Car Emission Standards.*

provisions that create the potential for state variance and allows for testing of the matching hypothesis. These provisions allow discretion in the implementation of I/M programs. As in stationary source policy, the country is divided into 247 air quality control regions with the states responsible for ambient standards in the regions within their boundaries.

Existence of I/M programs

The existence of I/M programs was initiated by the states but greatly advanced by the national government. Some subnational governments did show initiative in this issue. By the time of the 1977 amendments, seven localities had voluntarily begun I/M programs. The programs are described briefly in table 4.2. The date listed is the year of the formulation of the I/M plan. As the table shows, however, during this period few penalties existed against auto manufacturers for inadequate performance of emission controls. Further, I/M plans met "strong opposition at the state level."[32] As a result of lenient penalties and few operating plans at the state level, the federal government became heavily involved in 1977. I/M programs were mandated for thirty states and the District of Columbia. The other twenty states convinced the EPA that they could reach ambient standards by 31 December 1982 and were thus not required by law to establish I/M programs.[33]

State receptivity to I/M plans has varied. State adoption of an I/M program is not completely dependent on federal initiative. This is immediately apparent in the fact that at least three states (New Jersey, Oregon, and Arizona) set up programs even before required by federal law. Other states have adopted I/M programs only at the point of the federal gun. Pennsylvania was forced to set up a program after federal impoundment of highway funds. However, the state legislature did not make the program particularly stringent.[34] California authorized I/M in six polluted areas in 1982 but only after the EPA imposed funding restrictions in 1980 due to an SIP that did not include I/M legislation.[35] Both states claimed that federal imposition of such procedures were violations of states' rights. Other states (Missouri, Maryland, Indiana, and North Carolina) experienced delays due to various problems, not the least of which were reluctant state policymakers.[36] This reluctance to establish tough programs has diminished in recent years, a time period roughly corresponding to the conceptual resurgence of the states.

Matching between need and response is high relative to the existence of I/M programs since their presence is often based on EPA determination of air quality problems. Table 4.3 shows the results of logit equations on whether or not states operated I/M programs in 1989. Two equations are used because of the high multicollinearity between the presence of ozone and personal income per capita. The correlation between these two variables should not be surprising. Wealthier people own more cars. More cars produce more ozone. Any measure for presence of polluters (in addition to pollution levels) would also suffer this collinearity problem. The federal presence is explicit and thus not modeled as a variable. The resultant equations are very simple. Both equations are significant at very high levels and both air quality and relative affluence are strong predictors of the presence of an I/M program. Again, this is not a surprising result. A more revealing analysis measures those aspects of I/M programs determined by state policymakers. This is done in the next section.

Variance within I/M programs

Even though EPA ultimately approves and prescribes some aspects of I/M programs, the states are given wide discretion in how programs are designed

Table 4.3 Determinants of Existence of State I/M Programs (dependent var. = 1 if
I/M program in 1989, 0 if not)

Independent Variables	With Ozone	With Affluence
constant	−1.17	−8.29*
	(0.81)	(3.09)
o3aq[a]	0.45*	—
	(0.15)	—
pipc[b]	—	0.75*
	—	(0.26)
piindex[c]	3.23	−1.71
	(3.64)	(3.03)
Chi square	23.88	21.10
sig level	0.005	0.005
prediction rate	80%	74%
*significant at the 0.05 level		

a. percent of state population living in excess ozone
b. personal income per capita
c. index of party competition and turnout

and implemented.[37] The states enjoy considerable discretion in determining
procedures (although most use tailpipe checks) and cutpoints for models and
years. Texas, for example, set up a program that did not check tailpipe
emissions but rather concentrated on checking for tampering and improper
maintenance. The EPA estimated that the maximum reduction with this
program would be around 13 percent, well below the 25 percent minimum,
and thus may need further reconsideration.[38] Nearly all I/M tests consist of
"short tests" of concentrations in the tailpipe at idle, two-speed idle, and
when the engine is under a load.[39] Cost factors can also vary. Some states
mandate ceilings on the repair expenses of varying amounts (usually between
$50 and $100) while others (Oregon) require motorists to pay whatever is
necessary to pass the test.[40] EPA has also encouraged states to develop
antitampering enforcement programs. By 1987 thirty-two such programs,
including five statewide, were in operation.[41]

Two important and easily quantifiable measures of I/M programs are time
intervals between inspections and the applicability of standards to old cars.
These are important details. Tests could be conducted only at long-time
intervals, thereby allowing poorly controlled cars to pollute the roadways for
months, even years. The importance of the age factor is evident in the fact
that up to 85 percent of auto pollution comes from the oldest 50 percent of
cars on the road.[42]

Table 4.4   Determinants of Annual I/M Requirement (dependent var. = I if annual or less, o if not)

Independent Variables	With Ozone	With Affluence
constant	−1.14	−5.01*
	(0.71)	(2.28)
o3aq[a]	0.26*	—
	(0.10)	—
pipc[b]	—	0.41*
	—	(0.18)
piindex[c]	2.56	−0.38
	(2.99)	(2.79)
Chi square	10.00	8.02
sig level	0.01	0.05
prediction rate	66%	68%
*significant at the 0.05 level		

a. percent of state population living in excess ozone
b. personal income per capita
c. index of party competition and turnout

Tests on these variables are displayed in tables 4.4 and 4.5. Again, two equations are shown for each dependent variable because of multicollinearity between ozone and personal income. Again, the need variable and the affluence variable are significant predictors of state decisions to use at least annual tests and to test old model (pre-1974) cars. Discretionary state behavior thus reinforces the pattern of responsiveness apparent in tests on the presence of I/M programs.

Running separate equations for need and affluence is admittedly an incomplete means of disentangling this correlation. Conceivably, the measure for ozone could be a proxy for relative affluence. Nevertheless, each equation is statistically strong on its own. The strongest state programs are at least instituted in the states where MS pollution is highest.

Overall performance of the states

These results strongly support the matching hypothesis. Matching of need and response is readily apparent in MS pollution control. That matching is largely determined by high federal intervention and low interstate competition. Even when states exercise discretion in the details of their programs, stronger programs are present in states with higher MS pollution.

Table 4.5 Determinants of I/M Programs for Old Cars (dependent var. = 1 if pre-1974 cars tested, 0 if not)

Independent Variables	With Ozone	With Affluence
constant	−1.66*	−5.11*
	(0.76)	(2.16)
o3aq[a]	0.24*	—
	(0.09)	—
pipc[b]	—	0.37*
	—	(0.17)
piindex[c]	−0.19	−2.66
	(3.00)	(3.46)
Chi square	11.14	9.78
sig level	0.005	0.01
prediction rate	66%	66%
*significant at the 0.05 level		

a. percent of state population living in excess ozone
b. personal income per capita
c. index of party competition and turnout

Leading State Behavior in Mobile Source Air Pollution Control

Leadership by state governments in MS air pollution control policy centers around the behavior of the state of California. Data from the preceding analysis help identify California as a leading state. The state has an effective I/M program and is scored 10 out of 10 for air programs in general by the FREE authorities.[43] Historically, the California state MS program provided a model for federal policymakers in early national air pollution legislation.[44] Today, California initiatives still provide examples for other states and the federal government to emulate. This section describes the California program, with some discussion of current plans for Los Angeles, and the state's leadership role in this policy area.

Description

The California MS pollution control program is an important product of "America's most competent state government."[45] Competence in governing is essential to protecting the image of the good life that California strives to maintain. The stringency of policies in automobile controls is almost surprising considering the impositions placed on the millions of Californians who live by the car. Roughly 24 million cars and light trucks crowd the freeways of this state, seemingly all operating at all times.[46] Although critics of state governments might predict that all these mobile voters would prevent strict controls on their vehicles, California's air programs are cited as superlative even by officials from other state governments.[47] This section considers several aspects of the California MS program.

Organizationally, the state MS program is somewhat different from that concerning stationary sources of air pollution. Many of the SS decisions are made at the level of California's forty-four county and regional air quality agencies. Although some aspects of the MS program are also decentralized, most of the major policies are established in Sacramento by the California Air Resources Board (CARB). CARB consists of nine politically appointed members with a staff headed by two deputy executive officers. The board meets twice a month and is assisted by an Advisory Board on Mobile Sources, which includes representatives of fuel industries, auto manufacturers, universities, and the public community.[48] The board has access to a research program that is "second in size only to the federal EPA."[49]

CARB members are proud to term their antipollution standards as "among the toughest in the world."[50] California tailpipe emission standards are exempt from federal controls as long as they are at least as stringent as those applied nationally. Standards are quite detailed. Categories include cars, different classes of trucks, and motorcycles. Within those categories, emissions are specified by type of pollutant, miles driven, and year of vehicle. For example, the standards for a 1980 passenger car at 50,000 miles allow 0.41 grams/mile of HC, 9.0 grams/mile of CO, and 1.0 grams/mile of NOx.[51] In addition to tailpipe standards, California imposes ambient standards that are tougher than those promulgated federally. For example, the California standard for fine particulate matter is tougher than federal ambient standards.[52] Further, no federal standard for visibility exists, whereas California has imposed a standard of ten miles (when humidity is less than 70 percent).[53]

California's I/M effort is directed at areas of serious MS pollution. The initial reluctance to adopt I/M programs mentioned earlier has been replaced by a vigorous state effort in recent years. Programs were begun in the big cities of Los Angeles, Sacramento, San Diego, San Francisco, and San Jose in 1984.[54] Numerous other cities and counties now also conduct tests. Because these activities have developed at different times in different communities, the program is somewhat decentralized, but CARB does have considerable control over local efforts. Smog checks are enforced through vehicle registration. Further, all cars are checked when first purchased or first registered in California. This includes vehicles up to twenty years old, although recent changes affect cars that date back to 1966.[55] The smog tests always include a tamper test, which is enforced by allowing $50 limits for repairs unless the antismog equipment has been altered.[56] Smog Check certificates cost $6 while the average inspection fee is $21.[57] This money is used to defray the costs of the program.

Since 1984 the state claims to have reduced harmful auto emissions by 17 percent through its I/M program.[58] This progress is due to the stringent tailpipe standards and to the diligence of the testing. According to 1987 EPA data, California had the highest failure rates (25 percent) of any of the more than twenty states audited.[59] When a car fails the I/M test, it must be modified to reduce its emissions. Thus high failure rates reflect diligent testing. As the EPA states, the "primary reason for the low failure rates (in other states) was found to be improper testing."[60] One official within CARB admitted that the tough I/M program has inspired some opposition by the California AAA but that theirs was the "only voice in the woods" speaking against strict enforcement. The state's goal is to achieve 25 percent reductions in tailpipe emissions through its I/M program by 1994.[61]

Any discussion of mobile source controls in California without consideration of the peculiar problems facing Los Angeles would be incomplete. The air pollution problem is particularly severe in L.A. because of the huge number of drivers and because of the local geography. Thus, according to Gene Fisher of the South Coast Air Quality Management District (AQMD), even though the average pounds of pollution per person may be less in L.A. than in New York or Chicago, the surrounding mountains and the warm temperatures concentrate and exaggerate the air pollution problem. The ugly brown cloud seen from Mount Whitney triggers smog alerts eighty to ninety-five times per year.[62] Officials estimate that 70 percent of Los Angeles smog is caused by mobile sources.[63] The problem has been so traditionally persistent that air pollution control in the United States actually started in L.A. with the establishment of an air pollution district in 1947.

The Los Angeles basin is now subject to controls imposed by AQMD and CARB. AQMD consists of roughly 1,100 employees with an annual budget of $101 million. AQMD is an integral part of the state effort to control MS pollution. According to officials at both levels, the state and the district work closely together to develop policies and procedures. Not only does AQMD implement statewide rules but it also initiates efforts of its own. If AQMD (or any other regional office, for that matter) falters, CARB can take moderate steps, such as withholding gas tax dollars, to encourage behavior or can even resort to more extreme actions such as taking over the program.

Following complaints by environmentalists about slow progress toward air quality goals in the mid-1980s, AQMD asked for and received authority in 1987 to make changes in local life-styles as well as in smokestacks. The governing board of AQMD has developed an unprecedented plan (AQMP) for attacking air pollution that includes controls on everything from oil refineries to lawnmowers, utilities to barbecues. The plan was adopted on 17 March 1989 by AQMD and the Southern California Association of Governments and then approved by CARB in August.[64] AQMP provides for three tiers of effort as determined by available technology. The plan provides few specifics on costs and benefits but cost estimates for compliance run nearly as high as the promises of progress.[65] For example, the plan did suggest that Tier I controls could cost each resident in the basin up to sixty-five cents per day.[66]

Implementation of the plan is to take place in stages. Tier I controls utilize existing technology to address such mundane sources of pollution as aerosol sprays, underarm deodorants, and paints as well as refineries, public plants, and tailpipe emissions. In terms of mobile sources, Tier I controls require clean fuels for fleet vehicles, improve traffic flow and public transit, and impose control measures on diesel-powered vehicles, aircraft, locomotives, and pleasure boats.[67] Tier II requires converting 40 percent of passenger vehicles to clean fuel within ten years and reducing other MS emissions by 50 percent.[68] Tier III actions on mobile sources depend on developments "in fuel cells, solar cells, storage batteries, and superconductors (which) offer the promise of eliminating combustion processes from motor vehicles almost entirely."[69]

By 1990, AQMD had made significant progress. According to the annual progress report, many of the Tier I controls had been established.[70] In July the city proudly opened a subway line between Long Beach and downtown. Further, as of this writing, the AQMD was considering the possibility of commuter fees on freeway drivers to encourage transit usage.[71] Obviously, the most effective control of MS pollution would be to eliminate the mobile sources. In a city with geographical characteristics like those in Los Angeles, serious reduction in usage of gas-powered automobiles is essential for continued progress.

Leadership

California's leadership in MS pollution control policy is unquestioned. Federal policymakers have acknowledged this since 1970 by exempting the state from federal standards and allowing other states to emulate California's efforts. As described before in this book, leadership consists of willingness to change the status quo and the ability to attract followers. California MS policymakers have demonstrated an abundance of both.

California tailpipe emission standards often exceed those established at the federal level. California has set standards which "have generally led the rest of the country by from 2 to 5 years".[72] For example, currently, California has tougher standards than the federal government in the control of carbon monoxide emissions from cars.[73] In such cases, CARB must request a waiver from the EPA and be ready to show that the state standard is neither arbitrary nor capricious and that some commonality exists between federal and state standards. As an illustration, one CARB member suggested that a waiver would be approved if the state applied standards for 100,000 miles that the EPA used for 50,000, but that the waiver would be denied if the state said those miles had to have been driven clockwise when the EPA mandated them counterclockwise. The board claims that the state and federal staffs work closely together on new regulations. A good working relationship is apparent in the fact that the state has never been denied a waiver by the EPA. Still, one official did admit that state standards were beginning to differ more and more from federal limits.

California standards are emulated at both the federal and the state levels. The 1990 revisions of the Clean Air Act are illustrative of the former. California exhaust emission standards for light-duty vehicles mandate phase-in of vehicle compliance with nonmethane HC and NOx limits. Specifically, 40 percent of 1993 and 80 percent of 1994 cars must meet standards of 0.25 grams/mile NMHC and 0.40 grams/mile of NOx.[74] These were the exact levels proposed in both the House and Senate versions of revisions to the Clean Air Act.[75] Once again, the federal government is following California's lead in the control of tailpipe emissions.

An example of state emulation occurred recently with standards on gasoline volatility. Whereas the EPA mandated reduction of hydrocarbon evaporation in summer months to 9 psi by 1992, California has limited volatility to 9 psi since 1975 and is currently planning reductions to 8 psi.[76] Following California's lead in 1987, eight northeastern states formed a regional agreement to reduce gasoline volatility sold between 1 May and 15 September to 9 psi.[77] Seven of the eight states implemented those regulations in 1989, fully three years before the federal standard was to be lowered to this level.[78] Since states may only exceed federal standards in MS policy to copy California and even then only with explicit federal approval, the northeastern states submitted the required revised SIPs to EPA. As of mid-1989, five state revisions had been approved and the others were under review. Other states, including Pennsylvania, Delaware, Maryland, and Virginia, are also considering lowering volatility standards.[79]

Innovation, and emulation of innovation, in this policy area is not limited to standards. First, California continues to be a "testing ground for new automotive emissions control technology."[80] This technology is then available elsewhere throughout the world. Second, California, and particularly AQMD, experiment with process. For example, AQMD has recently begun a program that allows citizens to complain about smoking vehicles by calling 1–800-CUT-SMOG. Complaints may inspire warnings and citations.[81] Finally, independent actions by California may inspire actions by national manufacturers. State mandates for cleaner fuels in L.A. prompted a quick response by one gasoline producer. Arco announced that leaded gasoline at California pumps would be replaced by a reformulated, less-polluting fuel. The fuel will significantly cut emissions from pre-1975 vehicles, which currently account for roughly 30 percent of automotive pollutions in southern California. A spokesman for the company admitted that the reformulation was possible years before but only now, in response to the California initiatives and possible duplication by other locales, did the company have the incentive to produce the cleaner fuel.[82]

How does the process of emulation work? One means of information dissemination is through a nationwide organization. California is a member of the State and Territorial Air Pollution Program Administrators (STAPPA). STAPPA is a national association of state officials that recognizes the crucial role of state governments in air pollution control and therefore facilitates the exchange of information between agencies.[83] Another means of dispersion is through actual contact between representatives of different states. A CARB official acknowledged that other states have, although not that often, brought California officials to visit, testify, or provide guidance for their own efforts. Mostly, however, emulation works through imitation. The heavy involvement of the federal government in this policy area, including its adoption of California standards, facilitates the transmission of information, which allows for imitation to occur.

A different but related question concerns why leading states would contribute to the emulative efforts of others. When asked, California officials offered two reasons. First, they believe in their efforts as good policy. Second, they expressed awareness of the potential gap between their state's efforts and those of other states. If the state's standards and procedures get too far ahead of the rest of the country, national automakers might object to such diverse standards. After all, recall the receptivity of automakers to national standards in 1970. As a result, one administrator expressed concern that the federal government might feel pressured to "rein them in." Rather than holding back on their own actions, California can export its behavior and bring other states along with it.

Explanation

The independent behavior of California's MS policymakers is a reflection of the state's own relative detachment from the rest of the country. Politically, for example, California ranks as the state most independent of federal support for environmental programs by a considerable margin.[84] Politically detached by independence, geographically removed by mountains and deserts, sociologically separated by life-style attitudes, Californians thrive on the freedom to take care of themselves and to act upon their own needs and preferences.[85] These motivations provide the explanation for the strength of California's MS pollution control program.

One of the state's biggest needs, the need for effective pollution controls, is more than readily apparent. After responding with his "L.A." comment, the CARB official cited at the beginning of the chapter then added: "Californians have a high reliance on motor vehicles." Californians live by the automobile. As long as cars pollute, the need for strong programs will exist. Authorities predict that such usage will only increase. According to growth forecasts, population in the L.A. basin is expected to be 37 percent higher in 2010 than in 1985. More significantly, the total number of vehicle miles traveled will increase by nearly 68 percent.[86] Unless vehicle pollution is brought under control, the problem in L.A. and the rest of California will only get worse.

Another motivation is pride. When talking to CARB and AQMD officials, they voice awareness of being on the cutting edge. They know that other polluted areas are watching their efforts and judging California successes against their own problems. A CARB official admitted that one reason the state continues to push on MS control is that its efforts have been successful, more so than with stationary sources. Accomplishments breed continuing efforts. Further, AQMD official Fisher suggested that California's pollution problems today are tomorrow's crises in other big cities and added that we "either solve it in L.A. or the rest of the industrial areas are doomed." The awareness of being the nation's leader stimulates dedication in state officials and thus professionalism in state institutions.

Finally, the state's political culture is conducive to strong environmental efforts. One of the main reasons so many people move to California is because of the natural beauty of the state. That beauty is lost if it can only be seen as L.A. is from Mount Whitney. Not only are individuals aware of this danger, but so too are the recreation and tourism industries. Considerable pressure can thus be generated to encourage strong state efforts in these policy areas.

Conclusions

The dimensions of federalism governing mobile source air pollution control policy can be characterized as high in vertical involvement and low in horizontal competition. The federal government is heavily involved in mobile source control, particularly with the imposition of I/M programs on state efforts. Perhaps ironically, mobile sources of pollution cannot viably threaten massive relocation to other states with lax standards. Mobile source pollutants can cross state lines but not by design as emissions from tall smoke-stacks can.

The effects of these dimensions on state leadership are apparent in evidence supporting the three hypotheses of chapter 1.

Hypothesis 1. The lack of interstate competition and the heavy federal involvement in this policy area suggest strong matching of severity and state response. Indeed, states with higher levels of MS air pollution are more likely not only to have I/M programs (table 4.3) but also to have more stringent I/M programs (tables 4.4 and 4.5).

Hypothesis 2. Low potential horizontal interaction is conducive to leading state willingness to innovate and exceed federal guidelines. California is explicitly allowed supersedure. Never has the federal government denied the state a request for a new standard. Other actions within the state, such as the AQMP, are unprecedented at federal or state levels.

Hypothesis 3. The high degree of federal involvement in this policy area enables rapid communication and dissemination of leading state behavior. Other states have willingly copied California's standards, such as those in gasoline volatility. All states may be forced to copy tougher tailpipe standards if the federal government again emulates California's behavior.[87]

Cities like Los Angeles still suffer from mobile source air pollution. However, state policymakers have shown the willingness to innovate, emulate, and supersede federal guidelines in MS pollution control. Future success in improving the situation in the nation's cities is dependent on their continued efforts.

5 Nonpoint Source Water Pollution Control

A degree of irony accompanied Jim Gulliford's testimony before Congress in 1988. For two decades, environmentalists had spoken before the national legislature to seek relief from unresponsive state bureaucrats. Now, here was the director of the Soil Conservation Division for the state of Iowa testifying that the federal government was slowing his state's efforts to control nonpoint source pollution because those efforts were too stringent. What made the testimony truly ironic was that Iowa, with its heavy reliance on the agricultural community, was precisely the kind of state that many would have expected to be reluctant to address this issue in previous years.

This chapter considers the behavior of bureaucrats like Gulliford and states like Iowa in this policy area. The behavior of state governments has been somewhat responsive but widely variant. Federal involvement is so low that states act rather autonomously. One hypothesis thus is that dissemination of state responsiveness across state lines is minimal. A second hypothesis results from the low transfer of NPS pollutants and polluters across state lines. State officials are less afraid of relocation than in other policy areas and thus leading states are more likely to exceed even explicit federal standards.

The Role of the States in Nonpoint Source Water Pollution Control

Nonpoint source (NPS) pollution is diffuse runoff of harmful materials resulting from activities such as agriculture, mining, and forestry. Whereas point source discharges can be narrowly defined, NPS pollution emanates from wide areas of contamination. Contaminants include sediment, fertilizers, pesticides, oil, metals, acids, and debris. Although the need for policy response was recognized in 1970, little real intervention was forthcoming. For reasons described shortly, federal policymakers were and remain more interested in point source than in NPS pollution control. Vertical involvement in this policy area remains low. Further, the nature of the sources of these contaminants keeps the potential for horizontal competition low. Both dimensions are discussed below.

Low vertical involvement

The NPS problem has, since 1970, been recognized as serious but has not received federal attention commensurate to its severity.[1] Prior to the Environmental Decade, most land-use and related pollution control policies were the dominion of state and local governments. Pollution from land-use operations went virtually uncontrolled. The Federal Water Pollution Control Amendments (FWPCA) of 1972 authorized EPA redress of the problem in Section 208 but provided virtually no guidance to specifics. Section 208(b)(2)(F) mandates state plans for areawide waste treatment management, which are to include identification of NPS problems, establishment of methods of control, schedules of implementation, and provisions for continuous planning. Section 208 funding in the subsequent decade reflects the low priority given to NPS pollution. While Congress appropriated nearly $30 billion for waste treatment of point source discharge, the total appropriation for NPS pollution control through Section 208 was roughly $250 million.[2] As one eastern state NPS administrator told me, "They passed 208 and then dropped it on us."

Federal oversight has also been minimal. Although the EPA was given the authority to review NPS control plans, the agency has granted the states extremely wide discretion in their formulation. The agency limited its involvement in most instances to encouragement of voluntary application of Best Management Practices (BMPs) rather than explicit regulations of polluting behavior.[3] The basic BMP procedure involves general, systematic steps to analyze the NPS problem and then selection of the most suitable and cost-effective solution.[4] BMPs in agriculture, for instance, prescribe crop rotation to reduce erosion, reduction in pesticides, judicious use of fertilizers, and other recommendations specific to each site. Reliance on voluntary application of BMPs provides little means for federal intervention. Coordination of state NPS activities by the EPA has been more prominent of late but provides little more than recommendations and advice.[5] High-ranking officials recognize this negligence. In 1988, the EPA solicited opinions from seventy-five senior staffers on priorities for the agency. A vast majority of rankings placed NPS pollution in the category for high danger and low response.[6]

Agricultural practices obviously have a major impact on NPS pollution in many states. Thus direction from the U.S. Department of Agriculture (USDA), particularly in the area of soil conservation, is also a significant form of potential federal intervention. The department conducts several programs on its own to address environmental issues, such as the Agriculture Conservation Program, the Forestry Incentives Program, and the Water Bank Program.[7] USDA policy is based on two principles, voluntarism and localism. Most USDA programs are traditionally characterized by voluntarism so that behavior is only altered through "education, persuasion and financial incentives."[8] For example, by the end of the 1970s, the USDA budget for conservation was running roughly $1 billion, with nearly 40 percent of it devoted to cost-sharing.[9] The biggest portion of the remaining budget was spent on human resource management such as the Youth Conservation Corps, which also provides voluntary help. Program implementation is based on localized service, often through agencies as low as the county level. Decentralization ensures that national goals may take a back seat to local priorities.[10] As far as federal intervention goes, then, the abatement of NPS pollution resulting from agricultural conservation practices is thus dependent on voluntary cooperation by farmers. This voluntarism may not be insignificant, but it also provides no guarantees. For instance, while conservation tillage has been increasing in recent years, realistic assessments predict only half of the nation's cropland being used in this manner by the year 2010.[11]

Establishment of the Rural Clean Water Program (RCWP) by congressional amendment in 1977 could have had a significant impact on federal involvement in NPS pollution.[12] This program was created to provide long-term redress of agriculturally caused problems in heavily polluted areas. Projects consist of three- to ten-year contracts with volunteer farmers whereby the federal government provides technical assistance and cost sharing for up to 75 percent of the land user's expenses, up to $50,000 per participant. Contracted farmers were to install BMPs developed at the local level by the county Agricultural Conservation and Stabilization Committee and then approved by the state, the secretary of agriculture, and the EPA. Projects were initially selected within states by governors and State Rural Clean Water Coordinating Committees and then submitted to the federal government for approval.[13] While the program showed considerable promise, analysts even at the time suggested that low initial amounts of funding would minimize national impact.[14] More will be said on the results of some of the twenty-one approved projects later in the chapter.

The latest legislation affecting NPS water pollution control may provide some shift toward a greater role for the federal government in this policy. Changes in the role of the states have been formulated, as discussed below, in both agricultural and pollution control legislation.

The 1985 Farm Bill contained two elements that could have serious impact on agricultural pollution.[15] First, the "sodbuster" and "swampbuster" provisions discouraged planting on delicate farm areas or wetlands by denying federal benefits to farmers who tilled these areas. Second, the conservation reserve program was designed to take up to 45 million acres out of production in coming years by allowing the USDA to offer cash contracts to farmers who desisted from eroding practices on fragile land.[16] Consistent with other aspects of NPS control, the impact of these provisions will depend on state enforcement and implementation.[17]

The 1987 reauthorization of the Clean Water Act also addressed NPS pollution. The legislation required states to submit detailed reports discussing future redress of NPS problems (Sec. 316). The reports were to specify NPS controls necessary to achieve quality targets, the programs implementing the controls, and sources of funding.[18] The state reports were to be submitted within eighteen months, accepted or rejected by EPA in six months, and then, if rejected, resubmitted in three months. If the states failed to submit reports, then the EPA was authorized to write the plans for them. Federal grants up to 60 percent of the costs for state management programs were to be made available to approved states. State groundwater programs were also eligible for federal aid to the extent of $150,000 per year. These provisions received little real backing in the bill, however. Of the $18 billion price tag for the entire water legislation, only $400 million was designated to NPS control for the following four fiscal years.

The relatively low priority of NPS, as opposed to point source, pollution to federal policymakers is easy to understand in political terms. Point source pollution control provides opportunities for amelioration of highly visible, publicized discharge as well as the pork barrel of waste treatment plant construction. Federal redress of NPS pollution, on the other hand, conveys limited promise of quick, tangible benefits. Effective NPS policy also necessarily entails tangling with a powerful, cohesive, and well-entrenched agricultural sector, which is quite willing to debate government encroachments on the sanctity of traditional patterns of water usage.[19]

Many argue that the national government should take a more active role in NPS pollution control. State and local control of NPS pollution should, according to these arguments, be increasingly replaced, or at least supplemented, by national intervention. This shift would require greater financial involvement by the federal government, centralization of data collection and research, less discretion in state determination of standards and methods, and ultimately less variance between the states in their programs. Nationalization of NPS policy is advocated by many environmentalists, some agricultural economists,[20] observers,[21] analysts,[22] and influential federal entities.[23]

Even if political incentives should change, several reasons suggest that the protection of groundwater and surface water from NPS pollution is likely to remain a state prerogative. First, land and water use controls are subject to well-established systems of statutory and common law within each state. Replacing these with federal directives would not be easy. Second, important factors in NPS contamination such as hydrology and geology vary from state to state. Third, solutions to NPS problems often involve land use controls that require local knowledge of "institutional and political forces affecting their land resources."[24] Fourth, state officials welcome increased federal financial assistance, but federal directives are an unwelcome intrusion on their autonomy.[25]

Low horizontal competition

The issue area of NPS pollution control involves less potential for interstate competition than in the point source areas. This does not mean that interstate externalities and relocation threats by polluters are not possible, but rather that their impact on state behavior is less consequential in nonpoint than in point source pollution. This section discusses the likelihood of interstate externalities, the potential for relocation threats, and several reasons why the impact of horizontal competition in this area is relatively low.

Interstate externalities are low because of both the medium and the sources of pollution. Nonpoint source contaminants foul the water, not the air. Bodies of water can carry pollutants across state lines although they are less likely to do so than are currents of air. Further, sources of NPS pollution are not wedded to fast-moving bodies of water as are point source water polluters. Thus much of the contamination from NPS activities ends up in lakes and streams rather than rivers. State policymakers cite few instances of contamination resulting from external NPS pollution.

Interstate transfer of polluters is a different matter. Assume that one state imposes stricter NPS controls than any other state. Those controls reduce pollution in that state but also impose certain costs on local polluters, farmers for example. As a result, the stringent state farmers either raise prices to cover the costs or reduce supply to cut expenses. With relatively inelastic national demand for farm produce, farmers in other states can either raise their prices (without accompanying cost increases) or increase their supply, gaining revenue through either course.[26] Net farm income in the stringent state may be reduced.[27] Since these farmers in less-stringent states enjoy additional economic benefits, the stringent state farmers might feel tempted to relocate to lax states.

However, several factors mitigate the potential impact of relocation threats in the NPS policy area. First, polluters are tied to the land by economic strings.

Farmers, miners, and foresters are active in their particular location because of the availability of natural resources. Even if those resources are available elsewhere, someone will utilize the existing ones. An Iowa state official told the story of a hog farmer who threatened to move to Missouri if a stringent feed-lot bill was passed in Iowa. The Iowa official told him to go ahead, because there would be other hog farmers who could still make a profit with the new law. The farmer moved but the policy was not affected. Second, many NPS polluters, especially farmers, are tied to their current location by historical strings. A farmer whose land has been passed down for generations is much less likely to pick up stakes than the owners of a factory or utility. Third, unlike stationary source polluters whose individual exit can impact the local economy, farmers and other NPS polluters would have to leave en masse to have an effect. Since collective voice is easier to establish than collective exit, NPS polluters are more likely to unite to pressure their own state to avoid such stringent controls than to threaten mass relocation.

The Theoretical Model in the Current Context

Relative progress

States have been left largely to their own devices in this policy area and the results have occasionally been disastrous. In 1983, 700 families were evacuated from their homes in Times Beach, Missouri, after the rain-swollen Merrimac River overflowed. Residents feared that the subsequent flooding contained dioxin, a potentially dangerous chemical that had been sprayed on roadways years before to control dust. In 1985 huge concentrations of selenium, residue from agricultural drainage, contaminated the Kesterson Reservoir in California. The contamination killed hundreds of birds and wildlife.[28] These and other incidents, such as the leaching of cyanide from open-pit gold mining into groundwater in Montana and Nevada, are manifestations of uncontrolled NPS pollution.

The NPS problem is more widespread and systematic than these anecdotes reveal. Most experts agree that a major obstacle to achieving clean water in the United States is NPS pollution. NPS pollutants constitute roughly one-half of the pollution load deposited in America's waters.[29] The largest contributor, by far, is agricultural activity.[30] Nearly three-quarters of the states admit that agricultural runoff affects at least 50 percent of the surface waters within their state.[31] The effects of NPS pollution on groundwater are also quite severe but only now being realized. One study of the effects of agriculture on groundwater reports that 30 percent of examined wells had unusually high levels of nitrate and 8 percent were seriously dangerous.[32]

Hypothetical impacts of state governments

The theoretical model developed in chapter 1 generates several hypotheses concerning the impact of state governments on nonpoint source water policy failures and possibilities. With federal and interstate influences both relatively low, the matching of severity and response is determined mainly by internal factors. Hypothesis 1 argues that some matching occurs, although its presence is influenced by other variables, such as available resources and political institutions. Without fear of massive polluter relocation, Hypothesis 2 predicts that leading states do not hesitate to exceed federal standards when they do exist. Hypothesis 3 suggests that the lack of federal involvement precludes means necessary for coordination and dissemination of leading state responses.

Overall State Behavior in NPS Water Pollution Control

Appraisals of overall state performance in this policy area are highly divergent. One senior EPA official told me that the states are cognizant of the serious nature of the NPS problem and are currently responding in kind.[33] On the other hand, the most comprehensive nongovernment study to date concludes that "Their (state governments) performance has been extremely variable, but most often not deserving of praise."[34] I suggest that both characterizations contain some accuracy. This is possible because of the wide variance between state efforts and because of developments that have occurred in the years that separate the two appraisals. Thus some states are developing responsive programs, but the fact is that the results have yet to be seen. This becomes apparent in a chronological examination of state behavior.

Early state behavior

Data on early state behavior in NPS control is rare, perhaps because of slow recognition of the severity of the problem. However, a Resources for the Future study provides analysis of state NPS planning at the end of the 1970s. The states used were chosen because of heavy farming activity and thus in need of policy redress. The results of the survey are synthesized in table 5.1.[35] One should notice the dominance of voluntary approaches and the low awareness of NPS problems in even heavily affected states. Variance between the surveyed states is low, particularly in terms of effectiveness. In a related study, Crosson found some variance between states in terms of amount of

Table 5.1   Early State NPS Programs

State	Pressing Problem	208 plan	Appraisal
Iowa	Soil loss	Voluntary	High awareness, low effect
Nebraska	Irrigation runoff	Voluntary	Low awareness, low effect
Texas	Erosion	Voluntary	Low awareness, low effect
Arkansas	Problem identification	Voluntary	Some awareness, low effect
Georgia	Sedimentation	Voluntary	Some awareness, low effect
California	Irrigation	Voluntary, regulatory	High awareness, some effect

Source: Crosson and Brubaker, *Resource and Environmental Effects of U.S. Agriculture.*

land being tilled using conservation practices, but he attributed little of that variance to state programs and more to agricultural factors.[36]

Federal funding through Section 208 in this period was not only low, it may also have been misdirected. The examination displayed in table 5.2 shows that federal aid was not contingent on the presence of the NPS problem, state resources, or the existence of environmental demand in each state. Section 208 allocations were almost entirely dependent on population of each state. Indeed, this should not have been a surprise. At the time of passage of the FWPCA, Congress vaguely mandated that states submit 208 requests based on "urban-industrial concentrations or other factors" with no mention of agricultural or mining operations.[37] Since NPS pollution is generated by activities (such as farming) that reflect low population densities, this reliance on population is an odd choice of criterion.

What might have been affecting state behavior in this period in the absence of meaningful federal influence? One possible answer is systems of water property rights. Two basic common law institutions have developed in the United States regarding water as a property right.[38] Eastern states utilize a riparian system whereby rights to water are established by contiguity to bodies of water. Because water is relatively abundant, contiguity includes parks and residences as well as farms and mines. In times of shortage, none has priority and all must reduce usage together. The seventeen western states, on the other hand, establish water property rights by an appropriative system of "beneficial use." Because water is relatively scarce, priority on water rights is ranked by time of first usage. In times of shortage, the junior holders must reduce consumption first.[39]

Table 5.2 Determinants of Federal NPS Grant Money (dependent var. = 208 funds to each state)

Independent Variables	All states	Western states	Non-Western states
constant	−879	1818	−1121
	(2537)	(1589)	(2494)
%river[a]	−1.45	49.62	−19.42
	(27.39)	(38.74)	(30.88)
pop[b]	1.14*	0.93*	1.44*
	(0.078)	(0.086)	(0.12)
pipc[c]	0.25	0.033	0.07
	(0.21)	(0.342)	(0.32)
SCpc[d]	−577.2	69.03	462
	(457.7)	(431.25)	(1036)
R square	0.85	0.95	0.86
N	50	17	33

*significant at the 0.01 level

Dependent variable = Total Section 208 dollars for FY 74–80
a. % of state's rivers affected by NPS pollution
b. population of each state in 1985
c. personal income per capita by state
d. Sierra Club members per capita by state

Conceivably, differing systems of property rights could determine the variance in NPS programs. The appropriative system of property rights is a major construct of the relationship between state/local government and NPS pollution producers in the western states. Since little concern for the environment existed in early years of western settlement, the "first come" users of water were largely farmers and miners. Western states are still subject to intensive agricultural and mining operations. Thus, if we assume that state governments are susceptible to internal economic pressure, western state policymakers are unlikely to forcibly alter existing land and water use behavior of these NPS polluters. No systematic data exists to enable testing of this proposition in the early years of NPS control, but it will be examined in the following section.

NPS control during the 1980s

Analyses of state NPS programs during the 1980s display considerable variance. Without describing all of the technical details, states made choices on both methods and goals of their NPS programs. In terms of methods, each source category of NPS pollution can be subjected to voluntary restraints, mandatory constraints, some combination of the two, or to no guidance whatsoever. In agricultural pollution programs, for instance, roughly half of the states utilize only voluntary programs while the other half use some combination of voluntary, regulatory, and cost-sharing methods.[40] Similarly, goals for state water programs display wide range. Groundwater standards, for example, differ between numeric specifications of maximum contaminant concentrations in twenty-six states and narrative prohibitions about general discharge in fifteen states. Numeric standards reflect ambient quality targets that can account for nonpoint as well as point sources of pollution.[41] However, in many cases, these states have simply substituted EPA drinking water standards for groundwater goals, a potentially unrealistic course since groundwater quality often exceeds current drinking water targets.[42]

Examination of the simple choice between voluntary and regulatory programs is revealing. The simplest operationalization of state NPS efforts, that depicted in table 5.3, is determined by a state's NPS agricultural program. The variable is assigned a value of 0 if the program is purely voluntary and 1 if the program uses some regulatory or cost-sharing aspects in addition to voluntary encouragement. This is a simple differentiation of the stringency of state policies which figures quite prominently in discussions of NPS programs. Most environmentalists and many agricultural economists favor shifting all state programs from voluntary to regulatory, some terming "mandatory controls a must."[43] Farmers, other NPS polluters, and their political representatives are quite resistant to precisely such a shift. As one representative of the USDA told me, "The only way farmers will accept the feds is with education and volunteerism."

Results of testing of the model are displayed in table 5.3. Testing was done by a simple unordered logit package. The independent variables were described in chapter 1. The three columns of numbers represent logit equations with three different measures for the potential influence of the agricultural community. The first column utilizes the most accurate indicator of the importance of the agricultural community (*fmgsp*) to a state's economy: farm sector contribution to gross state product. This variable is highly significant and the equation as a whole is also quite powerful. Further, the variable representing relative affluence (*pipc*) of the state is also significant. The second column uses a rougher measure of agricultural presence in a state, but even this measure of percentage of land devoted to farming (*farm*) is statistically significant. Finally, the third column uses a dummy variable to check for the effect of different systems of property rights on state agricultural programs. Neither this variable nor its accompanying equation are at all significant. This column's lack of significance suggests that merely dif-

Table 5.3 Logit Equations on State NPS Regulation (dependent var. = 0 if voluntary, 1 if more)

Independent Variables			
constant	−5.92*	−4.53	−1.98
	(2.73)	(2.48)	(2.01)
fmgsp[a]	0.50*	—	—
	(0.21)	—	—
farm[b]	—	0.05*	—
	—	(0.02)	—
dummy[c]	—	—	0.50
	—	—	(0.66)
pipc[d]	0.38*	0.22	0.11
	(0.19)	(0.17)	(0.15)
piindex[e]	1.50	2.51	3.62
	(3.17)	(3.28)	(3.88)
pc208[f]	−0.67	−0.07	−0.21
	(0.46)	(0.37)	(0.36)
degrees of freedom	4	4	4
Chi square	13.74	7.46	3.30
significance	>0.01	>0.1	no
*significant at the 0.05 level			

a. farm sector as % of gross state product
b. % state acreage devoted to agriculture
c. 1 if appropriative, 0 if riparian system
d. personal income per capita
e. index of party competition and turnout
f. dollars of federal 208 aid per capita

ferentiating states according to systems of property rights does not explain behavior in this period even if it may have in the 1970s and before.

This analysis of the use of regulatory NPS programs indicates two things. First, states develop variance on an important dimension, the choice of regulatory mechanisms. Second, the variance between states on this dimension is determined significantly by need and affordability. This second result is considered again in the following section, but it should be mentioned at this point that additional testing of the strength of the agricultural variables was conducted. The first logit model was run twice more after deleting separately the variables for agricultural presence and relative affluence. In both instances, the equation was statistically different from the original equation with Chi-squares beyond the 0.05 level, strong evidence for the power of those deleted variables.

Another means of examining state variance in NPS control in this time period is to discuss the different experiences under the RCWP program. In all, twenty-one projects were funded at a total cost of nearly $70 million. Over-all, the program showed some success. Several projects showed that NPS problems could be controlled and that careful selection of BMPs could en-hance the likelihood that benefits would exceed costs.[44] In addition, much information was gathered about NPS control processes, some mark of success for an experimental program.[45] Farmer receptivity was crucial and variable, but studies of individual programs did show that some farmers would coop-erate to the point of exceeding their share of the funding for the project. For instance, of $4.3 million spent to clean up Tillamook Bay in Oregon, farmers contributed $2.2 million.[46]

Table 5.4 lists the RCWP project states and some assessment of each pro-gram.[47] Because it was an experimental program, the projects were selected to represent different NPS needs.[48] As the table shows, state efforts varied greatly in terms of expenditures (costs) and outcomes (benefits and improve-ments). Many of the projects are unlikely to show net benefit results. In addition, several will not markedly improve the quality of the water. Others reflect very positive experiences for the states. Unfortunately for those net benefit states, the varying results precluded a strong enough consensus to support the RCWP program.

Current state efforts in NPS pollution control

What does an overall measure of current state NPS efforts show? The preced-ing discussion suggests variance in state programs with some support for Hypothesis 1. State choice of means of control displays significant influence by the need and affluence variables. Conceivably, need and resources contrib-ute to state experiences with the RCWP program. But these are not conclusive results.

I constructed a ranking of state policy responses from a variety of sources. State policymakers themselves offer appraisals of both existing programs and management plans.[49] Quantitative data is available on statutory ground-water control measures, state monitoring capabilities, and categorization of different type regulatory strategies.[50] These factors were combined in one index and then compared to another comprehensive ranking that appraises state NPS programs by scope of response in relation to the severity of the problem.[51] The two indexes, which correlated at an extremely high signifi-cance level, were then aggregated to produce a comprehensive ranking of state NPS policies. This ranking was then compared, as a check, to the simple

Table 5.4   Assessment of RCWP Projects

State	Pollution due to Agriculture	Water Quality Improvement Likely	Gross Benefits	Governmnt Costs	Likelihood of Net Benefits
Alabama	75%	Yes	Moderate	High	Low
Delaware	30	Yes	Moderate	Moderate	High
Florida	27	Yes	High	Moderate	High
Idaho	100	Yes	Low	High	Low
Illinois	94	No	Low	Moderate	Low
Iowa	100	Yes	Moderate	Low	High
Kansas	100	NA	Low	Moderate	Low
Louisiana	100	Yes	Low	High	Low
Maryland	100	Yes	Uncertain	High	Uncertain
Massachusetts	50	No	Low	Low	Low
Michigan	100	Yes	Uncertain	High	Uncertain
Minnesota	90	Yes	Moderate	Moderate	Moderate
Nebraska	100	Yes	Low/mod	Moderate	Uncertain
Oregon	70	Yes	High	High	High
Pennsylvania	100	No	Uncertain	Moderate	Uncertain
South Dakota	100	No	High	Moderate	High
Tennessee/ Kentucky	100	No	Low	High	Low
Utah	15	Yes	Uncertain	Low	Uncertain
Vermont	25	Yes	High	High	High
Virginia	88	Yes	Uncertain	Moderate	Uncertain
Wisconsin	60	Yes	Uncertain	High	Uncertain

Source: Piper et al., "Benefit and cost insights," pp. 206, 207.

decisions concerning voluntary and regulatory programs. The correlation was statistically highly significant.

Second, the ranking was used in a multiple regression equation as displayed in table 5.5. The dependent variable in this equation is the ranking of state NPS programs. Since econometric questions exist concerning the usage of bounded rankings (1 to 10), the second column provides a logit analysis of the same model, this time with the dependent variable reflecting either strong or weak NPS programs. Obviously, the model is not completely specified. For instance, the equation only includes the farm sector and not other NPS polluting sectors such as mining and forestry. Nor does the equation include potentially important political variables such as party of the governor. This accounts for the low R-square term. Nevertheless, the variables for agricultural presence and affluence are again statistically significant beyond the 0.05

Table 5.5   Determinants of State NPS Control Programs (dependent var. = ranking of ST NPS programs: 1-10 or 0-1)

Independent Variables	OLS Estimation	Logit
constant	−0.31	−7.40*
	(2.23)	(2.77)
npsriv[a]	−0.02	−0.03
	(0.02)	(0.20)
fmgsp[b]	0.22*	0.21*
	(0.10)	(0.11)
pipc[c]	0.42*	0.46*
	(0.17)	(0.20)
piindex[d]	1.73	2.98
	(2.98)	(3.40)
pc208[e]	−0.43	−0.38
	(0.39)	(0.41)
R square	0.2	
Chi square		13.54
significance level		0.05
*significant at the 0.05 level		

a. % of state rivers affected by NPS pollution
b. farm sector as % of gross state product
c. personal income per capita
d. index of party competition and turnout
e. dollars of federal 208 aid per capita

level. Further, the logit equation is statistically quite powerful, indicating that state variance in this area is significantly influenced by these important variables.

Third, the ranking is used to examine those states studied ten years before. Indeed, the current rankings reflect considerably more variance than existed before. Current data suggests that the states that Crosson studied have indeed changed their programs. On my scale of 1 to 10, 10 being the most comprehensive, the states in table 5.1 received scores of, respectively, 10, 7, 2, 4, 8, and 6. Iowa is perhaps the most interesting case. Crosson and Brubaker were quite critical of Iowa's program in their survey, terming the state's efforts, particularly in soil conservation, as "not very effective."[52] Ten years later, many observers describe Iowa as having a superlative NPS program. The FREE surveys, specifically, score Iowa a 4 out of 5 for the scope of their NPS activities and a 10 of 10 for their soil conservation program.[53]

When I asked several Iowa policymakers about this change, they agreed that the state effort had changed considerably in the last few years. Increased resources, commitment, and experience had all improved the program during the period when states were being "revitalized."

Overall performance of the states

States have become more responsive to the NPS problem in the last few years. Comparisons of all fifty states show that some of the states with the worst problems, particularly those with relative affluence, are also those that have developed the toughest programs.

Before considering the Iowa case in depth, mention should be made of two issues where states have exceeded federal guidelines and are independently developing innovative approaches that could affect NPS pollution control. First, one phenomenon is broadly known as organic farming. Currently consisting of less than 10 percent of American agricultural output, organic farming involves growing produce with less fertilizers and pesticides. Each farmer's smaller output is balanced with higher premiums on healthier food. In Europe, for instance, where organic farming is more popular, one survey suggests people will pay as much as 35 percent more for organic food.[54] The cost-effectiveness of this kind of farming is also enhanced by improvements in soil fertility. Fledgling efforts in several states already show promising signs of success.[55]

Second, states have begun consideration of systems of water pricing. Although not yet applied to NPS pollution, water rights systems created through federal reduction of discounts for users and voluntary transfer of rights have already been proposed.[56] Experimentation with such proposals is ideal for the state level, especially in light of this analysis, which suggests some responsiveness to need by policymakers.[57]

Leading State Behavior in NPS Water Pollution Control

Data from the preceding analysis facilitates identification of Iowa as one of the leading states in nonpoint source pollution control. Iowa developed an early NPS program (table 5.1), displays a strong regulatory component (table 5.3), has experienced considerable success with RCWP efforts (table 5.4), and scores highest on the ranking used in table 5.5. Iowa is an interesting state to examine for NPS control. Iowa is "the preeminent Corn Belt state."[58] Given that more than 10 percent of the gross state product of Iowa comes from the agricultural sector (narrowly defined), it should not be surprising that NPS pollution has always been a serious problem. What may be surprising is that Iowa has developed one of the most effective NPS programs in the country. This section describes the program, with attention to its supersedure of federal guidelines, and then offers some explanations for its development.

Description

NPS pollution in Iowa results almost entirely from agricultural activities. According to the latest data, all assessed stream miles in the state were impacted "to some degree" by agriculture. In nearly all cases, agriculture was also the primary NPS contributor.[59] The potential for agricultural contamination is high in both surface and ground water. Pollutants include sediment, nitrogen fertilizers, herbicides, and pesticides. Among other unnatural sources of NPS pollution, only urban runoff constitutes a meaningful impact and it is still much less than that from agriculture.[60]

Iowans have been aware of NPS problems for several decades. Rules controlling pollution from animal feeding operations predate 1972 federal intervention by three years. The official state program directed at agricultural pollution control began in 1975. In response to Section 208 of the FWPCA, Iowa began several studies to determine the severity and extent of NPS pollution within the state. Recommendations of the studies were incorporated into the 1979 Water Quality Management Plan. The early program concentrated on soil conservation and sediment control.[61] State officials relied on traditional conservation measures and voluntary participation.[62]

Since 1979 the state has expanded the scope and stringency of its NPS program. The rules on animal feedlots have been revised several times until they now exceed EPA regulations.[63] Funding for various aspects of the management plan has been increased and numerous BMP projects have been conducted. Significant other programs and pieces of legislation have been added in recent years, including the Groundwater Protection Act of 1987, Iowa Soil 2000, and the management plan for meeting Section 319 requirements. The organizational apparatus overseeing these programs was solidified in 1986 with the consolidation of all environmental programs in the Department of Natural Resources. Implementation of NPS programs is mainly the responsibility of the Surface and Groundwater Protection Bureau and the Soil Conservation Division. These programs provide education, regulations, and funding in ways described below.

Education is the cornerstone of Iowa's NPS activities. State policymakers believe that educating polluters of the need for responsible practices is the most efficient and effective way to control NPS contamination. As one administrator explained, if you can get people to do things on their own, then the state doesn't have to go out to each individual site all the time. Further, many feel that most Iowa farmers are receptive to training and advice, especially if convinced of future benefits. Statewide education efforts are conducted in several ways. The Soil Conservation Division relies on its nearly 500 locally elected district representatives to provide technical assistance, soil surveys, data evaluations, and BMP recommendations to local farmers.[64] Since these reps are usually farmers themselves, their advice is often well received. In addition, the Cooperative Extension Service of Iowa State University provides education, information, and training programs throughout the state and is now currently responsible for most of the training of fertilizer and pesticide usage.[65] Further, the Integrated Farm Management Demonstration Project was established in 1986 to conduct demonstration activities on different plots with different technologies.[66] The Extension Service has been so effective that other states are now asking Iowa for advice on funding and administering such a program.

Still, regulation is utilized in some circumstances. Recently proposed rules call for limits on the use of atrazine, the most frequently detected agricultural chemical in groundwater. The rules would require that farmers in affected areas could apply no more than 1.5 pounds per acre per year.[67] Fertilizer and pesticide dealers are required to have impermeable storage areas and mixers. Farmers, if they use quantities of a certain size, are also subject to these regulations. Commercial applicators of pesticides must pay a fee and pass a test to be certified. Farmers applying pesticides are required to follow federal and state regulations for usage.[68] Above the level of individual farmers, each soil district must meet approved soil loss limits set by the State Soil Conservation Committee.[69] State administrators are proud of their combination of voluntary compliance backed by regulatory capacity. Administrators from two different programs both told me that while they did not feel that they had to use regulation that often, they would not hesitate when necessary.

Iowa has ensured that adequate funding is available for NPS control activities. Much of the funding comes from fees on polluting activities, such as the manufacture and distribution of fertilizers and pesticides. Other money comes from state lottery revenues.[70] The money is used for administration, cost-sharing, and interest-free loans. Cost-sharing of up to 75 percent is available for conservation programs in watersheds above public lakes and for soil conservation practices by landowners. The money must be made available before compliance is forced on the individual farmer.[71] The state legislature has also appropriated funds for interest-free loans to farmers using conservation practices.[72] Further, the state has utilized the Federal Land and Water Conservation Fund to purchase threatened wetlands. Finally, one projection for money available through the Resource Enhancement and Protection Program in coming years suggested that it may total nearly $30 million per year for conservation efforts.

Leadership

Iowa has displayed innovative efforts in various ways. State policymakers have considered experimentation in creative methods such as a ban on fall plowing and the use of buffer zones between farming activity and waterways. One administrator sent copies of Aldo Leopold's *Sand County Almanac* to his field officers to remind them of conservation needs. The state government has shown initiative in watershed programs. Three RCWP projects stand out. The state spent nearly $7 million on the Big Spring Groundwater Demonstration Project to get farmers in a drainage area flowing into a sinkhole to cooperate on conservation efforts. At the Prairie Rose and Green Valley projects, the state achieved BMP participation levels amongst landowners above 75 percent within three or four years for each.[73] An independent assessment of these projects suggests that the state, as a result of its efforts, will likely achieve improvements in water quality and positive net economic benefits.[74] Few of the assessed RCWP projects in other states received such a positive evaluation.

Leadership behavior by Iowa is particularly apparent in the willingness to formulate NPS standards more stringent than those stated by the federal government. Most state administrators think that the state is ahead of the federal government in these policy areas. Evidence for that viewpoint was offered by Soil Conservation director Gulliford in testimony before the U.S. Congress on implementation of the 1985 Farm Bill. Briefly, Iowa had established tolerable soil loss limits on erodible lands and had achieved extensive cooperation when the federal government mandated acceptance of certain noncompliant behavior. Gulliford testified: "This action creates immediate problems in that it significantly reduces the conservation accomplishments we can expect from the program, it creates confusion on the part of farmers, and it undermines the credibility of the (Iowa) conservation partnership."[75]

Recent developments in the soil loss controversy are particularly illustrative. The controversy had been simmering for years. Despite the federal foot dragging that inspired Gulliford's testimony, the state set plans to cut steep slope soil loss limits from 100 to 20 tons per acre per year. Gulliford characterized the federal response to me this way: "Right now, the federal government is more afraid of what we're doing than supportive." That fear was made manifest after a small group of Iowa farmers complained to federal Soil Conservation Service (SCS) officials that state limits would cut produce yields by financially harmful amounts.[76] Receptive federal authorities termed the Iowa state actions as "unduly inflexible" and called for relaxation of the regulations by doubling allowable losses to 40 tons/acre/year.[77] The SCS later added that all old soil loss regulations, including the 20 ton/acre limit, would be replaced by new ones following public meetings in each county.[78] The SCS actions inspired editorial denouncements in the *Des Moines Register*,[79] angry letters from conservation groups,[80] complaints from state officials,[81] and finally a confrontational meeting between U.S. SCS chief Scaling and members of the Iowa agricultural community where one farmer demanded Scaling's resignation.[82] A strong majority of the Iowa farm community supported the state officials. One survey of farmers showed 60 percent denouncing the original complaints on the 20 ton/acre regulation by agreeing that state limits would "minimally, if at all, adversely affect yields in their districts."[83] Even the Iowa Farm Bureau Federation (IFBF) said that most members did not want the rules relaxed.[84]

The controversy only ended when the federal government backed down. On 6 June 1990 the U.S. SCS announced that they would neither relax state limits nor hold county meetings.[85] On 11 July SCS chief Scaling was fired, having "rankled some farmers, professional conservationists, and environmental groups over enforcement of soil erosion standards."[86] Perhaps the bottom line on this story as it concerns the willingness of state leaders to exceed federal guidelines is found in a letter from the NRDC to Chief Scaling dated 30 May 1990. The letter demands that "Future revisions to plans should also align the federal program with state efforts."[87] In an interesting shift from the 1970s, environmentalists are demanding federal adherence to state standards.

Further evidence for state supersedure of federal standards is provided by an interest group that sometimes finds itself arrayed against environmentalists. While a representative of the IFBF admitted that their philosophy is generally based on the belief that "the closest government to you is the best," he added that was not the case in groundwater pollution control. As the IFBF *Resolutions* for 1990 assert, "state groundwater quality standards should not place agricultural activities nor other Iowa industries in a disadvantageous competitive position."[88] Thus, complementing the environmental shift mentioned above, polluters have shifted their efforts to more receptive ears at the federal level in NPS policy.

Diffusion of state efforts across state lines is rare. Two different Iowa officials admitted to me that they really did not communicate that much with other state NPS officials. In fact, as one suggested, the greatest improvement in the federal role would be for the national government to provide a clearinghouse for information so that states could benefit from each other's experiences. The federal government has shown little inclination in that direction. Experience in specific programs retains an emphasis on individual projects to the neglect of overall learning. The federal administrators of the RCWP admitted that the program was project-oriented to the point that little coordination across projects occurred. Further, as one suggested, indications were that Congress would move NPS policy even further in the direction of specific, individual projects. Renewal, as this offical said, will occur only in individual areas such as specific watersheds. One can easily imagine that members of Congress prefer such an alternative with its potential for credit-claiming as opposed to letting a federal agency select sites and coordinate state efforts in the future. Another analysis suggested that the RCWP successes that have occurred would need to be replicated at least 600 times to alleviate watershed problems throughout the country, an effort that would require billions of dollars.[89] Federal commitment of that kind of resources to facilitate dissemination of state RCWP achievements is unlikely.

Explanation

An event occurred while I was awaiting an interview in Des Moines that suggests reasons for the extensiveness of the Iowa NPS control program. I began a conversation with another person waiting in the lobby. He turned out to be a soybean farmer in town to get his pesticide license. I asked him if that was a nuisance, fully expecting a critical diatribe about bureaucracies. He responded, instead, "They sent me some stuff on it, there's something to it." Two factors suggested by this story, the initiating behavior of political institutions and the receptivity of the polluting sector, suggest reasons for Iowa's leadership in this policy area.

Iowa's political climate is conducive to aggressive political institutions. The state's political culture is an interesting combination of economic conservatism, international dovishness, emphasis on education, and belief in progressive reform.[90] The seemingly contradictory emphases on government intervention and traditional values are thus reconciled with Iowa's emphasis on education and voluntary compliance. The spirit of reform has stimulated strong, bipartisan environmental consciousness within the state over a long period of time. According to one top administrator, this consciousness pervades the agencies responsible for relevant policies. This is recognized and accepted by the agricultural community. One IFBF member described their interaction with the Iowa DNR as a "good working relationship" based on recognition of the different roles to be carried out. This characterization was echoed by an administrator who described credibility for both sides of the NPS issue based on the old philosophy that "good fences make good neighbors."

The political institutions themselves are strong and capable. Iowa's legislature meets annually for a limited period. Interestingly, one official suggested that the lack of a full-time legislature was probably a boon to progressive policies. Many legislators were also farmers and thus could provide both expertise and credibility to policy proposals. Several of these legislators, some with powerful positions, were mentioned by both administrators and lobbyists as having been quite aggressive in the NPS area. Bureaucratic agencies have credibility with elected officials and with interest groups. In fact, the previously mentioned Resource Enhancement Program was pushed by a powerful coalition of environmental agencies and environmental interest groups. Further, important bureaucrats enjoy direct lines of communication with elected officials and other relevant policymakers. Finally, one other factor is that resources are available for program implementation. State administrators seemed confident in continuing availability of resources. One suggested that even the discrepancy in federal funding between point and nonpoint sources had been balanced out somewhat due to passage of Section 319. That availability was not deemed to be contingent on changes in elected personnel. As one administrator told me, the political climate had more impact than who was in the governor's office did.

A second major reason for Iowa's aggressive NPS program is receptivity to policy efforts based on recognition of need. NPS administrators in Iowa are realistic about their achievements. They are modest about the high ratings that the state has received from outside sources and offer that they have not yet achieved the results they seek. They are quite proud, however, of the programs that have been instituted. Similar to responses that I have received in other states with effective programs, their simplest explanation for the extent of the program is that the severity of the problem was recognized and addressed. This recognition is not limited to actors within political institutions. According to a survey of residents taken in conjunction with the 1987 Groundwater Protection Act, 83 percent of respondents stated that "they felt more needs to be done to solve ground water pollution problems."[91] That recognition extends to the polluting sector. The IFBF refers to its members as "working environmentalists." Farmers are among the first to recognize the need for pollution control, especially since so many of them use ground water for drinking water. According to one representative, the IFBF, while still critical of some state programs, was ahead of both the federal Farm Bureau Federation and other state bureaus in the acceptance of some mandatory controls precisely because state members recognized the need for government response.

To summarize, the case study findings are consistent with the quantitative analysis in the preceding section. Iowa has developed a strong NPS control program because of the severity of the problem, a receptive political culture, capable institutions, and available resources. Neither the federal government nor private interest groups have a significant impact on policy efforts. In fact, though the IFBF says that it pressures state officials by reminding them that "Iowa is not the only state where you can grow corn," the federation has turned to the federal government for relief from stringent state efforts. Vertical influence is so low in this policy area, however, that the Iowa NPS program is what continues to grow.

Conclusions

The dimensions of federalism governing nonpoint source water pollution control policy can be characterized as low in vertical involvement and low in horizontal competition. Federal influence by the EPA and the USDA is limited by traditions of voluntarism and localism. Polluters and pollutants are less likely than in point source policy to cross state lines largely because of the natural resources that are being used or affected.

The effects of these dimensions on state behavior are apparent in the evidence supporting the three hypotheses of chapter 1.

Hypothesis 1. States are quite variable in their responsiveness to NPS pollution, but some correlation exists between severity of the problem and the response of the state government. As shown in tables 5.3 and 5.5, this correlation is more likely when states have relatively abundant resources.

Hypothesis 2. Leading state officials in NPS policy, at least in the case of Iowa, have shown ready willingness to innovate, experiment, and exceed federal directives. These efforts are apparent in regulatory mechanisms, RCWP projects (table 5.4), and the tolerable soil loss controversy that shook the federal SCS.

Hypothesis 3. The low level, and occasionally counterpurposes, of federal involvement in this policy area make dissemination of leading state efforts difficult. For example, the federal government provided anything but a means for dissemination of Iowa's efforts to reduce soil erosion losses to other states.

Slow overall progress in the abatement of NPS pollution will likely continue in the near future. State administrators are the first to admit that their efforts need more coordination. Even the EPA, which had until recently sugarcoated the NPS situation, has recently admitted that "nonpoint pollution is a major remaining water quality problem which will prevent the achievement of established water quality goals, even when applicable point source controls have been fully implemented."[92] Compare that admission to this EPA assessment two years earlier: "EPA can report that a significant amount of activity and resources is being devoted to identifying and controlling nonpoint source pollution problems at the Federal, State, and local levels of government."[93] Nevertheless, the willingness of some states to exceed minimal federal guidelines suggests that federal policymakers should be careful about how they intervene in this pressing situation.

Two hundred years ago, James Madison warned of the behavior of state governments with the following account:

> It is no longer doubted that a unanimous and punctual obedience of 13 independent bodies, to the acts of the federal Government ought not to be calculated on.... How indeed could it be otherwise? In the first place, Every general act of the Union must necessarily bear unequally hard on some particular member or members of it, secondly the partiality of the members to their own interests and rights, a partiality which will be fostered by the courtiers of popularity, will naturally exaggerate the inequality where it exists, and even suspect it where it has no existence, thirdly a distrust of the voluntary compliance of each other may prevent the compliance of any, although it should be the latent disposition of all.[1]

Madison foresaw the fears of state parochialism which pervade the formation of many domestic policies even today. Until recently, Madison's fears appeared warranted at least in the area of environmental policy. The record showed a "historic lack of attention and occasional resistance to natural resource and pollution control programs shown by state officials."[2] Nevertheless, in the early 1970s American politicians gave state policymakers ultimate responsibility for the implementation and efficacy of pollution control policies. Many argue that within the last few years, state governments have taken the lead in these policy areas.[3] The extent, depth, and impact of that leadership has been the subject of this book.

The thesis of this work is that state leadership is affected by what I call the dimensions of federalism: the vertical dimension of federal involvement in state behavior and the horizontal dimension of interstate competition within each policy area. This chapter reviews that argument and offers cross-policy perspectives that could not be considered in the previous chapters, which dealt with single policy areas. First, I summarize the model of state leadership in pollution control policies, discuss the dependent variable and its importance, and recount the independent variables often used to explain variation in state policy outputs. Then I reconsider the dimensions of federalism and how they differ between policies. Second, I will recapitulate the findings on the specific hypotheses that were generated by the model in chapter 1. Comparative results between policy areas will be made explicit. Finally, I briefly speculate on the generalizability of the model to pollution control policies in the future, other environmental issues, and other domestic policies.

A Model of State Leadership in Pollution Control Policies

State leadership in pollution control policies

The dependent variable in this study has been and will continue to be important. Prior to the policy developments discussed in the individual chapters, governmental impact on environmental issues almost completely depended on state and local policymakers. The significant increase of federal involvement in the early 1970s did not diminish the overall importance of state behavior. Rather, state governments were assigned new responsibilities, which varied between policies. The 1980s brought increased authority to state governments in nearly all environmental policies as the Reagan administration delegated and decentralized.[4] The Bush presidency has shown little inclination to change this reliance on state governments, nor can centralization be expected in times of federal budget deficits and general criticism of big government. In the 1990s more people will look to the state governments for leadership in these policy areas. If air and water pollution control are important to quality of life, then the leadership of state policymakers is crucial.

Are state governments providing the leadership necessary for effective pollution control in the United States? No simple answer to this question exists. Subsequent sections of this chapter will be more explicit, but obviously state leadership is indeed a variable, not a yes or no proposition. This book has used a model with the following components to assess the consistency and the exceptionalism of state leadership.

Potential determinants of variation between states

In chapter 1, I utilized extant literature to develop a model of variation between state governments. State governments are arguably affected by the severity of policy needs, the pressure of relevant interest groups, the availability of resources, political culture within the state, and federal aid. These variables were combined into one model and used in numerous assessments of the behavior of all fifty states. The model itself is generally fairly powerful, as reflected by the significance levels of most of the equations. Each variable displays importance at least once, although measures for political institutions are rarely significant. On the other hand, the affluence variable (*pipc*) is often strong, lending considerable support for the logical argument that a state's behavior is often influenced by the availability of resources.

This model of variation between states warrants utilization in its current form but also further expansion. It can tell us much about the consistency of state performance within specific policy areas and has been the subject of much recent literature.[5] This book builds upon that literature by expanding the model across policy lines. To understand the leadership of states in policy efforts, specific policies warrant differentiation and comparison. Indeed, pollution control policies differ in important ways. These differences have received considerably less attention in the environmental policy literature.[6] This book has argued that such differentiation is necessary in order to understand the extent, depth, and impact of state leadership. I placed the preceding model within two theoretical dimensions of federalism.

Vertical involvement

Policies differ along the vertical dimension of intergovernmental relations. Federal policymakers have assigned state governments different responsibilities in different policy areas. Further, the federal government maintains a greater presence in some policy areas than in others. Of the four policies examined, the federal presence is greatest in point source water and mobile source air pollution control. In both cases, the federal government has been involved in specifying means of attainment for state governments: NPDES standards and POTWs in the former, inspection and maintenance programs in the latter. Vertical involvement has been lower in stationary source air and nonpoint source water pollution control policies. In both cases, the federal government has set attainment goals for the states (ambient standards in air, Section 208 guidelines in water) but has left methods and procedures largely up to state policymakers.

Horizontal competition

Policies also differ along the horizontal dimension of interstate competition. States can compete with each other over both polluters and pollution that can cross state lines. Of the four policies examined, horizontal competition is greatest in stationary source air and point source water pollution control. Polluters can viably threaten relocation as single entities and can release their pollutants into wide-ranging media (high levels of the atmosphere for air, strong rivers for water). Horizontal competition is lower for mobile source air and nonpoint source water pollution control policies. In both cases, polluters are too numerous and diverse to threaten relocation collectively. Further, they release their pollutants in ways that are less likely to cross state boundaries than are pollutants from their stationary, point source counterparts.

The accounts in the case studies therefore verify placement of the pollution control policies in the cells of figure 1. Indeed, policies can be differentiated according to dimensions of federalism. What does that tell us about state leadership?

Performance of the Model

The model of state leadership within the dimensions of federalism was tested with three hypotheses. The results are summarized comparatively in this section.

Hypothesis 1. Some correlation between severity and response is apparent at the state level, but overall matching varies according to the dimensions of federalism:
a. when vertical involvement is low and horizontal competition is high, matching is less apparent;
b. when vertical involvement is high and horizontal competition is low, matching is more apparent;
c. when vertical involvement and horizontal competition are both high or low, matching depends on other variables.

This hypothesis considers the extent of state leadership. State governments can provide national leadership with consistent policy responsiveness to policy needs at the state level. Indeed, one of the most important debates in the current literature on federalism concerns the relative impact of the needs variable in models of state government behavior. According to the state resurgence argument, the impact of policy needs on policy responses at the state level may become more noticeable as states develop responsive institutions. Evidence to date on this question is and continues to be mixed. Some correlation of response to presence of pollution is apparent in all four policy areas (tables 2.7, 3.4, 4.3, and 5.5). However, the findings in the case studies also suggest that the appropriateness of response varies across policy lines in ways that reflect the hypothesis stated above.

Evidence on the three qualifications of Hypothesis 1 is telling. In stationary source air policy, where vertical involvement is low and horizontal competition is high, state governments exhibit widely variant behavior. The needs variable is less deterministic in substantive policy responses (tables 2.7 and 2.8) than it is for symbolic outputs like intentions (table 2.4) and expenditures (table 2.5). Quite the opposite occurs when vertical involvement is high and horizontal competition is low, as in mobile source air pollution control. States suffering from greater severity of pollution are likely to have the strongest responses (tables 4.4 and 4.5). In the two areas of water pollution control, when the dimensions of federalism are in balance, the likelihood of matching is significantly affected by the availability of resources to the state government (tables 3.7, 5.3, and 5.5).

The most obvious conclusion from the apparent variation in state matching behavior is that not all states are responding appropriately to policy needs within their borders. For strong programs to be apparent, weak programs must also exist. If matching between need and response were always high and weak programs existed only where pollution was low, this would not be a problem. However, this is not the case. Indeed, the four states examined closely score highly in most, including overall, environmental scales. For example, in its 1987 summary, FREE ranked Wisconsin, California, North Carolina, and Iowa as 1, 2, 6, and 8 out of the fifty states.[7] Some states never make the top ten in any specific policy area. These potential "pollution havens" can affect not only the local residents but also the behavior of other state governments. One cannot assert that the resurgence of the states described in chapter 1 has occurred across the board.

These results generate several other conclusions. First, support for the hypothesis suggests that the dimensions of federalism do have an impact on the likelihood of matching at the state level. Second, when the dimensions of federalism are not unbalanced enough to sway the model in either direction (toward or away from matching), then responsive behavior by state governments is more likely to occur when it is affordable. Third, as other analysts have warned, drawing conclusions about state behavior based on symbolic responses such as expenditures may be unrealistic.[8]

Hypothesis 2. The lower the level of interstate competition in a policy area, the more likely that leading state programs supersede federal guidelines.

This hypothesis examines the depth of state leadership. For state leaders to be important, state policymakers have to show the willingness to innovate, experiment, and go beyond what higher authorities require of them. Innovations and creativity were apparent in all four case studies. As the hypothesis suggests, however, the potential for interstate interaction affects the depth of those developments. Innovations in stationary source air (e.g., Wisconsin market incentives) and point source water (e.g., North Carolina basin permitting) are state modifications on existing federal practices rather than dramatic supersedure of federal standards as in mobile source air (e.g., California tailpipe emission controls) or nonpoint source water (e.g., Iowa soil loss regulations).

Another interesting finding on this hypothesis stems directly from the interstate dimension and relates to the discussion of horizontal competition in chapter 1. Many of the innovations at the state level are realized only after exhausting political fights. Wisconsin's air toxics rule, North Carolina's aquatic toxicology program, Iowa's soil loss regulations, and Los Angeles's air quality plan all inspired serious opposition. As the interstate differentiation of these policies would suggest, however, only in the former two did the opposition use viable threats of exit. In Iowa and California, opponents of the plans relied on voice to try to alter or modify the proposals.

Hypothesis 3. The higher the degree of vertical involvement in a policy area, the greater the dissemination and coordination of leading state efforts.

This hypothesis concerns the impact of state leadership. For state leaders to have an impact on overall state behavior, other states must follow their lead. The development of other state programs can include either the diffusion of leading efforts or the coordination of interstate behavior. The degree of federal involvement is a key determinant of such development. The federal government now plays an active role in the pollution prevention program begun in North Carolina and continues to force other states to adopt California's emission standards by copying them in federal legislation. On the other hand, in policies where vertical involvement is low, coordination of state efforts is problematic. Despite regional proposals like Wisconsin's, amelioration of acid rain control awaits federal involvement in state efforts.[9] States such as Iowa continue to act autonomously in the control of nonpoint source pollution, and in fact, federal coordination through programs like the RCWP seems to be diminishing.

Analysis of this hypothesis also sheds some light on a theoretical discussion in chapter 1. Policies with heavy vertical involvement allow state leaders to expand the scope of "conflict" on specific issues by nurturing their relationship with the federal government. California's consistent anticipation of federal reactions to waiver requests and North Carolina's provision of information on pollution prevention facilitate federal consideration of innovations that then become national in scope. These experiences provide sharp contrast to the lack of federal involvement in Wisconsin's air toxics proposals and the icy federal reception to Iowa's attempts to institute soil loss regulations. Toxic air pollutants and erodible lands remain issues contested in narrow arenas.

Broader Implications for the Model

How generalizable are these results? The model of state government leadership that includes attention to dimensions of federalism offers important findings on pollution control policies. Can it provide insight into future developments in these areas, other environmental issues, or even other domestic policies? I offer some speculation on these questions in this section. These conjectures are offered not as answers but as questions for possible future research.

Future pollution control efforts

To consider possible future developments in these areas, I discuss three sets of important actors in the model. First, many state policymakers have the knowledge and the motivation to lead public efforts. As described numerous times in the case studies, policymakers in leading states know the impact of variables such as federal influence and horizontal competition on their leadership behavior. In terms of motivation, I believe that many leading state policymakers seek to disseminate their ideas solely for the sake of effective public policy. But even if that did not suffice, policymakers would be motivated, as California air administrators attested, by recognition of the need to narrow the gap between strong and weak states. As a result, I expect to see more efforts by state policymakers to address the impacts from federal dimensions. These may include attempts to diminish the impacts of negative interstate competition, Wisconsin's regional acid rain settlement for example, and more utilization of vertical involvement through solicitation of federal assistance in interstate seminars such as North Carolina's Pollution Prevention conference.

Second, public good interest groups also recognize the potential impacts of dimensions of federalism on pollution control efforts. The leadership activity of state governments is becoming increasingly apparent to more environmental groups. I anticipate more application of environmental efforts at state and local levels than may have been the case twenty or even ten years ago.[10] For instance, the Greens party is trying to build strictly on a local basis in the United States.[11] Nevertheless, environmentalists do recognize that in some issues, the impact of horizontal competition on state efforts precludes systematic state responsiveness. In those instances, environmental groups may then concentrate their pressure on changing the vertical dimension of the policy. This seems to have been the case with the latest version of Clean Air legislation. Previously reluctant to specify means of attainment in stationary source control, federal policymakers, under considerable pressure from environmental groups, have taken the dramatic step of regulating specific power plants in order to abate acid rain.

Third, business and industry groups may also emphasize selective lobbying strategies based on federal dimensions. First, aware of the aggressive efforts of some state governments, business interests could concentrate more of their resources on federal policymakers in these instances. In particular, these groups will seek federal restraints on aggressive state programs, as has the Iowa Farm Bureau Federation. On the other hand, polluting groups will retain their focus on the state level in cases of undeveloped programs. Industrial groups will not easily relinquish pollution havens that may now exist, no matter how much potential influence state leaders exert. Depending on the success of polluter efforts, the variation between state programs could persist, if not actually increase. The combination of these factors suggests that state governments will be the battlegrounds for important pollution control issues in the years to come.

Other environmental issues

Theoretical impacts of the dimensions of federalism are apparent in other environmental policies. One important issue concerns the disposal of hazardous wastes.[12] Despite the Resource Conservation and Recovery Act of 1976 and the Superfund of 1980, meaningful vertical involvement in state government behavior has been low. Federal efforts have often been characterized as either "dilatory"[13] or even scandalous.[14] Occasionally, the EPA would take the lead in a cleanup, but often the federal government provided some funds and then let state governments develop the plans and do the work.[15] Horizontal competition over avoidance of hazardous waste deposits has been so pervasive that it has earned an acronym: NIMBY for Not In My Back Yard. States and localities have fought to avoid deposits of hazardous materials in their locales. Only in the last few years have states such as New Jersey, Minnesota, and California shown the leadership necessary for effective public response.[16] By 1987 twenty-nine states offered incentives to reduce or recycle hazardous wastes and twenty-nine states have Superfund statutes.[17] State leadership has prompted changes in the vertical dimension as the federal government now requires hazardous material inventories, emergency plans, and "cradle to grave" tracking of wastes.[18]

Another area of environmental policy where the dimensions of federalism conceivably affect state leadership is in the development of alternative energy sources. Vertical involvement in this policy area before the 1980s displayed an interesting mix of federal funds and state administered programs, such as the State Energy Conservation Program and the Solar Energy and Energy Conservation Bank.[19] States showed increasingly effective leadership with these programs until the drastic cuts of the Reagan administration in the early 1980s.[20] Now state policymakers act rather autonomously with low federal involvement. Some states such as California and Wisconsin continue to develop alternative sources of energy. Interestingly, the horizontal competition in this policy area may be of a more positive nature than exists in most other environmental issues. Presumably, states that develop energy efficiency without the costs, including externalities, of fossil fuels, can offer both cleaner living as well as cheaper energy in the future. Such progress may stimulate horizontal competition between states over creative developments.

Finally, although beyond the scope of this work, one could utilize theoretical dimensions of analysis similar to those herein to address two different levels of analysis. First, like state governments, local governments are also seriously affected by vertical involvement and horizontal competition. Promising work on environmental issues at the local level might incorporate these structural dimensions in future studies.[21] Second, altering the level of study in the opposite direction, global environmental policies are subject to serious influence from structural dimensions. Vertical involvement in these issues can take the form of United Nations resolutions, world conferences, or multi-country agreements such as the Montreal Protocol.[22] Horizontal competition obviously exists across international boundaries. National leadership on issues such as global warming and the protection of endangered species could be the subject of future studies using a model similar to the one developed in this book.

Other domestic policies

State leadership in numerous American domestic policies may be affected by the theoretical phenomena described in this book. Many social welfare policies, for example, involve low vertical involvement and high horizontal interaction. Thus inconsistent state leadership and high variance between state programs has resulted. For example, state governments make fundamental decisions on the basic needs of recipients of Aid to Families with Dependent Children. State payments varied in 1987 from $114 per month in Alabama to $533 per month in California.[23] State policymakers decide maximum monthly food stamp benefits with a resulting range of $308 per family of three in Hawaii to $117 in California.[24] State governments decide on eligibility for programs like Medicaid where the percentage of poor receiving coverage ranges from 20 to 90 percent.[25] In these and other domestic issues, the relocation of policy recipients leads to increased horizontal interaction between states. State governments may well compete in a downward cycle to avoid an influx of such beneficiaries.[26] Again, although systematic study falls beyond the scope of this work, application of a model that differentiates policies by federal dimensions may be revealing in these issues.

Conclusion

In great centralized nations the legislator is obliged to give a character of uniformity to the laws, which does not always suit the diversity of customs and of districts; as he takes no cognizance of special cases, he can only proceed upon general principles; and the population are obliged to conform to the requirements of the laws, since legislation cannot adapt itself to the exigencies and the customs of the population, which is a great cause of trouble and misery. This disadvantage does not exist in confederations; Congress regulates the principal measures of the national government, and all the details of the administration are reserved to the provincial legislators. One can hardly imagine how much this division of sovereignty contributes to the well-being of the the the states that compose the Union.[27]

This chapter began with a quote from Madison and finished with one from Tocqueville. These issues do not reflect new questions but rather renewed interest. The federal structure, which Madison viewed with such potential wariness and Tocqueville with such potential acclaim, still holds significance in affecting the behavior of important participants in the policy process. Pollution control policies in the United States vary according to the dimensions of federalism. The vertical involvement of the federal government and the horizontal competition from other states significantly affect the state leadership that is so crucial to progress in these policies. Future policy efforts may well depend on understanding the effects of these dimensions.

Notes

1. A Theory of State Leadership

1   Madison's most explicit condemnation of state modifications can be found in the manuscript "Vices of the political system of the U. States" as published in *The Writings of James Madison*, ed. Hunt, vol. 2, p. 364.

2   Perhaps the most damning criticism comes from William H. Riker, who suggested that "the main effect of federalism since the Civil War has been to perpetuate racism" ("Federalism," in *Handbook of Political Science*, ed. Greenstein and Polsby, vol. 5, p. 101).

3   Bowman and Kearney, *The Resurgence of the States*; Van Horn, ed., *The State of the States*.

4   Van Horn, "The Quiet Revolution," p. 1.

5   Bowman and Kearney, "Dimensions of State Government Capability," pp. 341–62.

6   Montesquieu, *The Spirit of the Laws*, p. 120.

7   Elazar, *American Federalism: A View from the States* (1966 ed.), p. 216.

8   Baum, "State Supreme Courts"; Rosenthal, "The Legislative Institution"; Van Horn, "The Quiet Revolution."

9   Bowman and Kearney, *The Resurgence of the States*, p. 40.

10   Van Horn, "The Entrepreneurial States," p. 211.

11   Gormley, "Custody Battles in State Administration," p. 135.

12   Lester, "A New Federalism," p. 59.

13   Nathan, "The Role of the States in American Federalism," p. 19.

14   Van Horn, "The Quiet Revolution," p. 5.

15   Gramlich, "Federalism and Federal Deficit Reduction"; Kraft, Clary, and Tobin, "The Impact of New Federalism on State Environmental Policy"; Lester, "A New Federalism."

16   Beyle, "From Governor to Governors," p. 36.

17   Rosenthal, *Legislative Life*; Rosenthal, "The Legislative Institution."

18   Baum, "State Supreme Courts."

19   Bowman and Kearney, *The Resurgence of the States*; Davis and Lester, "Federalism and Environmental Policy"; Garnett, *Reorganizing State Government*; Hebert and Wright, "State Administrators."

20   Game, "Controlling Air Pollution"; Lester et al., "Hazardous Waste, Politics and Public Policy"; Williams and Matheny, "Testing Theories of Social Regulation."

21 Lester and Bowman, "Implementing Environmental Policy"; Regens and Reams, "State Strategies for Regulating Groundwater Quality."

22 Bowman and Kearney, "Dimensions of State Government Capability."

23 For an interesting, recent argument related to this perspective, see Olson, *The Rise and Decline of Nations.*

24 Riker, "Federalism."

25 Kaufmann, *The Forest Ranger*; May, *Federalism amd Fiscal Adjustment*; McConnell, *Private Power and American Democracy*; Schattschneider, *The Semisovereign People.*

26 Inman, "Fiscal Allocations in a Federalist Economy"; McConnell, *Private Power and American Democracy*; Schattschneider, *The Semisovereign People.*

27 Lester and Bowman, "Implementing Environmental Policy"; Williams and Matheny, "Testing Theories of Social Regulation."

28 See Inman, "Fiscal Allocations in a Federalist Economy," for a review.

29 Dye (*Politics, Economics, and the Public*), in particular, has long argued that wealthy states are more likely to have strong programs than poorer states. Walker ("The Diffusion of Innovation among the American States") concurred in his work on innovation.

30 Dawson and Robinson, "Inter-Party Competition"; Game, "Controlling Air Pollution"; May, *Federalism and Fiscal Adjustment*; Wenner, "Enforcement of Water Pollution Control Law"; Williams and Matheny, "Testing Theories of Social Regulation."

31 Bowman and Kearney, "Dimensions of State Government Capability," p. 359.

32 Elazar, *American Federalism* (1984 ed.), p. 109.

33 Calvert, "The Social and Ideological Bases of Support"; Key, *Southern Politics.*

34 Peterson and Rom ("American Federalism, Welfare Policy, and Residential Choices") used an index of party competition and voter turnout with some effect in studying welfare policies.

35 Elazar, *American Federalism* (1984 ed.), pp. 117 and 136. Elazar's categorization was used with some success by Blomquist ("Exploring State Differences").

36 Calvert, "The Social and Ideological Bases of Support"; Calvert, "Party Politics and Environmental Policy"; Lester et al., "Hazardous Wastes, Politics and Public Policy."

37 Chubb, "Federalism and the Bias for Centralization"; Derthick, *The Influence of Federal Grants.*

38 Goggin et al., *Implementation Theory and Practice*, p. 150; Lester and Bowman, "Implementing Environmental Policy."

39 Davis and Lester, "Federalism and Environmental Policy"; see also Lester, "A New Federalism."

40 This should not be overstated. Nations can and do compete, but the subsequent behavior is modified by the degree of control that national governments have over the mobility of people and capital.

41 Chubb and Peterson, "American Political Institutions," p. 8.

42 Fiorina ("Legislative Choice of Regulatory Forms") is particularly noted for his "shift the responsibility" model.

43 Mayhew, *Congress: The Electoral Connection*, p. 135.

44 Though not used in this context, the most famous explanation of these concepts is by Hirschman (*Exit, Voice, and Loyalty*).

45 Tiebout, "A Pure Theory of Local Expenditures," p. 418.

46 Olson, "The Principle of Fiscal Equivalence," p. 483.

47 Lowi, "American Business."

48 Eberts and Grunberg, "Jurisdictional Homogeneity and the Tiebout Hypothesis"; Peterson and Rom, "American Federalism, Welfare Policy, and Residential Choice."

49 Stafford et al., *The Effects of Environmental Regulations on Industrial Location*; Tran, "Locational Factors."

50 Kieschnick, "Taxes and Growth"; Schmenner, *Making Business Location Decisions*.

51 Duerksen, *Environmental Regulation of Industrial Plant Siting*.

52 Stafford's conclusions are based on fifty-four interviews with and 104 questionnaires returned from business executives (Stafford et al., *The Effects of Environmental Regulations on Industrial Location*).

53 Schmenner's book (*Making Business Location Decisions*) is designed as a guide for executives faced with locational decisions.

54 Kieschnick admits that policies may be affected if "only a relatively small share of subsidized firms are actually influenced" ("Taxes and Growth," p. 215).

55 For example, a major part of Michael Dukakis's 1988 presidential platform was an invitation to look at the success of Massachusetts as evidence for his leadership.

56 Bluestone and Harrison, *The Deindustrialization of America*; Jacobs, *Bidding for Business*.

57 Specifically, thirty of thirty-two respondents agreed that competition has become more intense. Twenty-one of thirty-one said that because of increasing competition, they kept detailed data on their chief rivals, generally their closest neighbors (Jacobs, *Bidding for Business*, p. 7).

58 For several years, the CPA firm Alexander Grant and Co. has published versions of *General Manufacturing Business Climates* with regulations itemized and ranked by state stringency because they are considered important factors to consider for locational decisions.

59 Davis and Lester, "Federalism and Environmental Policy"; Rowland and Marz, "Gresham's Law."

60 The most famous exposition of this is in Federalist 10.

61 Elazar, *The American Partnership*; Grodzins, *Goals for Americans*; Scheiber, "The Condition of American Federalism."

62 Walker, "The Diffusion of Innovation among the American States," p. 890.

63 The most compelling examination of this phenomenon is found in Schattschneider (*The Semisovereign People*).

64 McConnell, *Private Power and American Democracy*.

65 For just two perspectives on capture, see Stigler, "The Theory of Economic Regulation," and Culhane, *Public Lands Politics*.

66 Kaufman, *The Forest Ranger*.

67 Olson, *The Logic of Collective Action*.

68 Moe, *The Organization of Interests*.

69 The most seminal pieces in the former category include Maass, "Divisions of Powers"; Grodzins, *Goals for Americans*; Elazar, *The American Partnership*; Riker, *Federalism*; Croy, "Federal Supersession"; and Hanus, "Authority Costs in Intergovernmental Relations." The latter category includes such insightful analyses as those by Derthick, *The Influence of Federal Grants*; Ingram, "Policy Implementation Through Bargaining"; Peterson, *City Limits*; and Chubb, "The Political Economy of Federalism."

70 Bowman and Kearney, *The Resurgence of the States*, p. 256.

71 Copeland, *Federal-State Relations in Transition*, p. 42; Lave, "Health, Safety, and Environmental Regulations," p. 138; Maxwell, *The Fiscal Impact of Federalism*.

72 Davies and Davies, *The Politics of Pollution*, pp. 44–47; Rohrer, *The Environment Crisis*, pp. 84–88.

73 Davies and Davies, *The Politics of Pollution*, p. 48.

74 U.S. Council on Environmental Quality, *Public Opinion on Environmental Quality*, p. 7.

75 Erskine, "The Polls," p. 121.

76 *Washington Post*, 23 April 1970, p. A1.

77 Nelson, now a consultant for the Wilderness Society, was interviewed in Washington, D.C., on 28 August 1986.

78 Davis and Lester, "Federalism and Environmental Policy"; Kraft, Clary, and Tobin, "The Impact of New Federalism"; Lester, "New Federalism and Environmental Policy."

79 Bowman and Kearney, *The Resurgence of the States*; Bowman and Kearney, "Dimensions of State Government Capability"; Davis and Lester, "Federalism and Environmental Policy"; Lester, "A New Federalism."

80 Ingram and Mann, "Preserving the Clean Water Act"; Lester et al., "Hazardous Wastes, Politics and Public Policy"; Regens and Reams, "State Strategies for Regulating Groundwater Quality"; Tobin, "Revising the Clean Air Act."

81 Brown and Garner, *Resource Guide to State Environmental Management*; Duerksen, *Environmental Regulation of Industrial Plant Siting*; Ridley, *The State of the States: 1987* and *The State of the States: 1988.*

82 Lester and Lombard, *The Comparative Analysis of State Environmental Policy.*

83 Goggin et al., *Implementation Theory and Practice*, p. 109.

84 Summary indicators are available in Duerksen, *Environmental Regulation of Industrial Plant Siting.*

85 Financial data is taken from U.S. Department of Commerce publications for various years; Brown and Garner, *Resource Guide to State Environmental Management*, for recent years; and unpublished U.S. Environmental Protection Agency (EPA) data.

86 FREE is also discussed in Lester, "A New Federalism," pp. 63–64.

87 Davis and Lester ("Federalism and Environmental Policy," p. 81) categorize North Carolina as low in capacity and high in dependence, Wisconsin as high in both, and California and Iowa as high in capacity and low in dependence. Further, these four states are widely spread on the need-response table cited in their chapter (pp. 78–79).

88 Walker, "The Diffusion of Innovation among the American States."

89 Crandall, *Controlling Industrial Pollution*; *Washington Post*, 4 June 1989.

90 Bowman and Kearney, *The Resurgence of the States*, p. 258.

2. Stationary Source Air Pollution Control

1 *United States Code* (U.S.C.) 42, & 7405.

2 U.S.C. 42, & 7401.

3 Miernyk and Sears, *Air Pollution Abatement and Regional Economic Development.*

4 Davies and Davies, *The Politics of Pollution*, p. 51.

5 Degler, *State Air Pollution*, p. 1.

6 Ibid., p. 2.

7 *Congressional Record*, 91st Cong., 2d sess., 21 September 1970, p. 32902.

8 Crandall, *Controlling Industrial Pollution*, p. 11.

9 As shown in table 2.2, Class A sources number 28,000 whereas NSPS sources number roughly 3,000. These data are taken from the 1988 EPA document *Progress in the Prevention and Control of Air Pollution in 1986.*

10 Melnick, *Regulation and the Court*, p. 44.

11 These are reviewed in Cheremisinoff and Morresi, *Air Pollution Sampling and Analysis Deskbook.*

12 Harrington and Krupnick, "Stationary Source Pollution Policy," p. 106.

13 *Congressional Record*, 91st Cong., 2d sess., 21 September 1970, p. 32918.

14 Manness and Rudzitis, "Federal Air Quality Legislation," p. 513.

15 *The Economist*, 22 August 1987.

16 Tobin, "Revising the Clean Air Act."

17 Toxics rules will be discussed later. For more on this increasingly-important issue, see the publication by the U.S. General Accounting Office (GAO), *States Assigned a Major Role in EPA's Air Toxics Strategy*.

18 Davies, "Environmental Institutions," p. 150.

19 *The Washington Post*, 13 June 1989, p. A1.

20 Vig, "Presidential Leadership," p. 52.

21 Kraft, "Environmental Gridlock," p. 113.

22 Since proposal, the Bush administration has already backed off from its alternative fuel plans.

23 Copeland, *Federal-State Relations in Transition*, p. 28.

24 U.S. GAO, *EPA Needs to Improve Its Oversight of Air Pollution Control Grant Expenditures*, p. 9.

25 This sudden reversal of implicit policy is described in a legislative history of the PSD policy contained in the *U.S. Code Congressional and Administrative News*, 1977, p. 1183.

26 As cited in Melnick, *Regulation and the Courts*, pp. 82–83.

27 *Congressional Quarterly Almanac 1976*, pp. 133, 143.

28 This provision was adopted by the House on an amendment by Breaux (D-Louisiana) on a vote of 237 to 172. A similar amendment was rejected in the Senate.

29 Melnick, *Regulation and the Courts*, p. 112.

30 Garvey and Streets, *In Pursuit of Clean Air*.

31 Section 126 of the CAA.

32 Vestigo, "Acid Rain and Tall Stack Regulation," p. 731.

33 Ibid., p. 738.

34 Boyle and Boyle, "Acid Rain," p. 26.

35 Ackerman and Hassler, *Clean Coal/Dirty Air*, pp. 44–54; Crandall, *Controlling Industrial Pollution*, chap. 7.

36 U.S. Congress, Senate Subcommittee on Environmental Pollution, *Enforcement of Environmental Regulations*, p. 142.

37 U.S. GAO, *EPA's Delegation of Responsibilities*, p. v.

38 U.S. EPA, *Progress in the Prevention and Control of Air Pollution in 1986*, p. VIII-1.

39 *Congressional Quarterly Almanac 1981*, p. 505.

40 Magazine, *Environmental Management in Local Government*; Melnick, *Regulation and the Courts*.

41 Ringquist ("Regulating State Air Quality") tested the correlation between the FREE rankings and reductions in sulfur and nitrogen oxides between the years 1973 and 1985.

42 *The Washington Post*, 4 June 1989, p. A18.

43 *The Washington Post*, 9 June 1989, p. A3.

44 Roberts and Farrell, *Approaches to Controlling Air Pollution*; Walker and Storper, "Erosion of the Clean Air Act."

45 Walker and Storper, "Erosion of the Clean Air Act," p. 203.

46 Melnick, *Regulation and the Courts*, p. 165.

47 Roberts and Farrell, *Approaches to Controlling Air Pollution*, p. 159.

48 Melnick, *Regulation and the Courts*, p. 220.

49 Corkin, "Comment," p. 189.

50 These are pollutants likely to be emitted by stationary sources. Automobile pollution uses a similar measure but for excess ozone. This data was presented in Crandall, *Controlling Industrial Pollution.*

51 Most of these data comes from the FREE study by Ridley on *The State of the States: 1987,* app. 1, pp. 8–12.

52 Much of this particular discussion is taken from the GAO report *States Assigned a Major Role in EPA's Air Toxics Strategy.*

53 U.S. GAO, *EPA Needs to Improve Its Oversight of Air Pollution Control Grant Expenditures.*

54 Data were collected from annual reports on individual state expenditures between 1970 and 1980 as published by the U.S. Bureau of the Census in *Environmental Quality Control* reports. Data reporting was terminated with the Reagan administration.

55 Game, "Controlling Air Pollution"; Sharkansky, "Government Expenditures and Public Services"; Williams and Matheny, "Testing Theories of Social Regulation."

56 Jones, "Regulating the Environment," p. 419.

57 U.S. EPA, *Compilation and Analysis of State Regulations.*

58 U.S. GAO, *Improvements Needed in Controlling Major Air Pollution Sources.*

59 U.S. EPA, *National Air Audit System,* p. I-1.

60 U.S. GAO, "Air Pollution: EPA's Inspections."

61 Russell et al., *Enforcing Pollution Control Laws.*

62 Ibid., p. 39.

63 Melnick, *Regulation and the Courts,* p. 220.

64 Vestigo, "Acid Rain and Tall Stack Regulation," p. 716.

65 Ibid., p. 717.

66 Smith, "Playing the Acid Rain Game," p. 268.

67 Vestigo, "Acid Rain and Tall Stack Regulation," p. 714.

68 Smith, "Playing the Acid Rain Game," p. 269.

69 Vestigo, "Acid Rain and Tall Stack Regulation," p. 730.

70 Melnick, *Regulation and the Courts,* pp. 227–231.

71 Smith, "Playing the Acid Rain Game," p. 279.

72 Further details of these and other related cases are available in Rogers and Petersen, "Air Pollution across State Boundaries."

73 Webber, "Equitably Reducing Transboundary Causes of Acid Rain," p. 228.

74 Martha Hamilton, "New Era Dawning for the Utilities," *The Washington Post,* 11 June 1990, Bus. sec., p. 1 and p. 30.

75 Barone and Ujifusa, *The Almanac of American Politics 1990,* p. 1312.

76 Renshaw, Trott, and Friedenberg, "Gross State Product," pp. 38–46.

77 Webber, "Equitably Reducing Transboundary Causes of Acid Rain," p. 228.

78 Jessup, *Guide to State Environmental Programs,* p. v.

79 Ridley, *The State of the States: 1987,* p. 4.

80 Ibid., app. 1, p. 3.

81 Wisconsin Department of Natural Resources (DNR), *Wisconsin Administrative Code,* chap. NR 400.

82 Jessup, *Guide to State Environmental Programs,* p. 489.

83 Wisconsin DNR, *Expanding Industry,* p. 2.

84 Jessup, *Guide to State Environmental Programs,* p. vi.

85 Wisconsin DNR, *Expanding Industry.*

86 Eighteen states spent more per capita on air programs in FY86 than Wisconsin did. If you

take an average for the decade of the seventies, twenty-four states spent more per capita than Wisconsin did. This provides evidence that expenditures are not always the best measure of state program effectiveness.

87 Wisconsin DNR, *Expanding Industry*, app. D.

88 Jessup, *Guide to State Environmental Programs*, p. 491.

89 See copies of *E E News* (published quarterly) and pamphlets such as "Educating the Decision Makers of Tomorrow" (1990) from the Wisconsin Association for Environmental Education.

90 Wisconsin DNR, *Wisconsin Administrative Code*, chap. NR 369.

91 The industrial coalition lost in circuit court but has appealed to the Wisconsin State Court of Appeals.

92 Wisconsin DNR, *Wisconsin Administrative Code*, chap. NR 396.

93 Council of State Governments, *The Book of the States 1984–85*, p. 345.

94 Barone and Ujifusa, *The Almanac of American Politics 1990*, p. 1312.

3. Point Source Water Pollution Control

1 Lake, Hanneman, and Oster, *Who Pays for Clean Water?*; Rosenbaum, *Environmental Politics and Policy*.

2 Ball, "Water Pollution and Compliance Decision Making."

3 Zwick and Benstock, *Water Wasteland*, p. 168.

4 Heath, *A Comparative Study*, pp. ii–iii. The six programs were Maine, Michigan, New York, North Carolina, Pennsylvania, and Virginia.

5 Luken and Pechan, *Water Pollution Control*, p. 2.

6 Lieber, *Federalism and Clean Waters*, p. 14.

7 Ackerman et al., *The Uncertain Search for Environmental Quality*.

8 To be more precise, BPT (best practicable technology) limitations are usually based on the most effective current performance among plants within an industry or subcategory of an industry. BAT (best available technology) limits are more stringent standards applied to entire categories of industrial dischargers. Since 1979, the EPA has promulgated twenty-seven final BAT regulations (*National Water Quality Inventory*, 1987, p. 112).

9 This approach has been used in some cases. For example, Wisconsin utilized WLAs to contribute to a 90 percent reduction in BOD (biochemical oxygen demand) due to pollution from paper mills and municipalities along the Wisconsin River (EPA, *National Water Quality Inventory*, 1987, p. 107). Cases such as these are the exceptions that prove the rule, however.

10 Strom, "Congressional Policy-Making."

11 Lieber, *Federalism and Clean Waters*, p. 11.

12 Copeland, *Federal-State Relations in Transition*, pp. 25–27.

13 Advisory Committee on Intergovernmental Relations (ACIR), *Protecting the Environment*, p. 44.

14 This impoundment led to the Budget and Impoundment Act of 1974 and a court decision demanding release of the funds in 1975.

15 Kovalic, *The Clean Water Act of 1987*, p. 11.

16 Ibid., p. 20.

17 Ibid., p. 25.

18 *Congressional Quarterly Almanac 1977*, pp. 706–7.

19 Kovalic, *The Clean Water Act of 1987*, p. 18.

20 Ingram and Mann, "Preserving the Clean Water Act."

21 Kovalic, *The Clean Water Act of 1987*, p. 28.

22 Davies, "Environmental Institutions," p. 150.

23 Kraft, "Environmental Gridlock," pp. 110–11.

24 Section 101(a)(3).

25 Kovalic, *The Clean Water Act of 1987*, p. 17.

26 U.S. GAO, *Serious Problems Confront Emerging Municipal Sludge Management Program*, p. 2.

27 U.S. EPA, "Privatization of Municipal Wastewater Treatment."

28 Goggin et al., *Implementation Theory and Practice*, p. 95.

29 U.S. GAO, *Serious Problems*, p. 11.

30 Esposito, "Air Pollution," p. 47.

31 Ibid.

32 U.S. Congress, Senate Committee on Public Welfare, *Hearings on Water Pollution Control Legislation*, pp. 70, 75.

33 Caldwell, *Man and the Environment*, p. 46.

34 Boudon, *The Unintended Consequences of Social Action*.

35 Opinion Research Corporation, "Public Opinion," p. 17.

36 U.S. CEQ, *Environmental Quality* (1973), p. 380.

37 These figures are derived from data reported in *CQ Weekly Report*, 6 August 1971, p. 1680, and 19 August 1972, p. 2067.

38 These figures are found in *CQ Weekly Report*, 4 November 1972, p. 2918, and 5 February 1971, p. 324.

39 U.S. Congress, Senate Committee, *Hearings on Water Pollution Control Legislation*, pp. 612, 665; U.S. Congress, House Committee on Public Works, *Hearings on Water Pollution Control Legislation*, pp. 1162, 1518.

40 Zwick and Benstock, *Water Wasteland*.

41 *NY Times*, 8 November 1971, p. 1, and 19 December 1971, p. 1.

42 U.S. Congress, Senate Committee, *Hearings on Water Pollution Control Legislation*, pp. 736, 754.

43 Lieber, *Federalism and Clean Waters*, p. 64.

44 Ibid., p. 35.

45 U.S. Congress, Senate Committee, *Hearings on Water Pollution Control Legislation*, p. 54.

46 The perspectives on this and other matters by Billings were obtained in an interview on 20 October 1987.

47 See the testimony of Senators Baker (R-Tennessee) on page 32921 and Senator Cooper (R-Kentucky) on page 32918 of *The Congressional Record*, 21 September 1970 and Senator Randolph on page 38805, 2 November 1971. Senator Cooper was interviewed in October 1987.

48 *Congressional Record*, 4 October 1972, p. 33693.

49 U.S. EPA, *National Water Quality Inventory* (1987).

50 U.S. GAO, *Stronger Enforcement Needed to Improve Compliance at Federal Facilities*, p. 11.

51 These authorities include the EPA, *National Water Quality Inventory*; the Association of State W.P.C. Administrators, *America's Clean Water*; U.S. EPA and U.S. Fish and Wildlife Service, *1982 National Fisheries Survey*; and the U.S. Geological Survey, *National Water Summary 1983*.

52 The former figure is from testimony by EPA assistant administrator Durning in 1979 (U.S. Congress, Senate Subcommittee on Environmental Pollution, *Enforcement of Environmental Regulations*, p. 142) while the latter is taken from a 1987 EPA analysis (*National Water Quality Inventory*, 1987).

53 Of 150 major federal facilities, roughly 20 percent are in noncompliance for any given quarter and 40 percent of those have exceeded limits for over a year (U.S. GAO, *Stronger Enforcement Needed*, p. 3).

54 U.S. Congress, Senate Subcommittee, *Enforcement of Environmental Regulations*; U.S. GAO, *Costly Wastewater Treatment Plants*.

55 U.S. EPA, *National Water Quality Inventory* (1987), p. 109.

56 See U.S. GAO, *Stronger Enforcement Needed*, for details on delegated authority as of December 1988.

57 Ridley, *The State of the States: 1988*, p. 44.

58 The dates for these approvals are summarized in Goggin et al., *Implementation Theory and Practice*, p. 53.

59 The lack of correlation between quality of water and strength of program could also result from the time between these measures, or it could reflect the improvement on water quality by those same programs. Nonpoint source pollution is the subject of chapter 5.

60 See also Freeman, "Air and Water Pollution Policy," on this point.

61 Russell, Harrington, and Vaughan, *Enforcing Pollution Control Laws*, p. 37.

62 Ridley, *The State of the States: 1988*, app. 1, p. 5.

63 U.S. GAO, *Stronger Enforcement Needed*, p. 62.

64 Ridley, *The State of the States: 1988*, app. 1, p. 4.

65 Ibid., p. 5.

66 Goggin et al., *Implementation Theory and Practice*, p. 53.

67 Ridley, *The State of the States: 1988*, app. 1, p. 8.

68 Ridley, *The State of the States: 1987*, app. 5, p. 3.

69 North Carolina Department of Environmental Management (DEM), *Water Quality Progress*, p. 4.

70 Ibid., p. 92.

71 Jessup terms the North Carolina permit program "pervasive" in *Guide to State Environmental Programs*, p. 344.

72 N.C. DEM, *Administrative Code*, sec. 15A NCAC 2H.0100, p. 5.

73 N.C. DEM, *Water Quality Progress*, p. 96.

74 Ibid., p. 93.

75 N.C. DEM, *Administrative Code*, sec. 15A NCAC 2H.0118, p. 13.

76 Ibid., 2B.0101.

77 N.C. DEM, *Water Quality Progress*, pp. 93, 108.

78 Ridley, *The State of the States: 1988*, app. 1, p. 9.

79 N.C. DEM, *Water Quality Progress*, pp. ii, 94.

80 Ibid., pp. i–ii.

81 Ibid., p. 93.

82 Ibid., p. i.

83 One exception may be the Champion paper mill which is the subject of a dispute between North Carolina and Tennessee.

84 N.C. DEM, *Water Quality Progress*, p. i.

85 N.C. *Administrative Code*, sec. 15 NCAC 2H.1000; N.C. DEM, *Water Quality Progress*, pp. 98–106.

86 N.C. DEM, *Water Quality Progress*, pp. 73–76.

87 N.C. DEM, *Organizational Chart*, 1 January 1990.

88 N.C. DEM, *Aquatic Toxicity Testing*, p. 3.

89 Ibid., p. 5.

90 N.C. DEM, *Water Quality Progress*, p. 112.

91 FWPCA, Sec. 101(a) prohibits "discharge of toxic pollutants in toxic amounts."

92 Ridley, *The State of the States: 1988*, app. 1, p. 16.

93 N.C. DEM, *Water Quality Progress*, pp. 112–13.

94 N.C. DEM, *Aquatic Toxicity Testing: Questions*, p. 14.

95 See Michael Royston's comments, among others, in the North Carolina Department of Environment, Health, and Natural Resources, *Pollution Prevention Pays: Symposium*, p. 3.

96 N.C. Department of Environment, Health, and Natural Resources, *Pollution Prevention Tips*.

97 See National Roundtable of State Waste Reduction Programs, *Summary*, for a summary of state efforts. See also U.S. EPA, *Minimization of Hazardous Waste*; U.S. Office of Technology Assessment, *Serious Reduction of Hazardous Waste*.

98 U.S. EPA, *Waste Minimization*.

99 National Roundtable, *Summary*.

100 U.S. GAO, *Serious Problems Confront Emerging Municipal Sludge Management Program*, pp. 15–16, 31.

101 Barone and Ujifusa, *Almanac of American Politics 1990*.

4. Mobile Source Air Pollution Control

1 White, *The Regulation of Air Pollutant Emissions from Motor Vehicles*, p. 5.

2 U.S. EPA, *National Air Quality and Emissions Trends Reports, 1987*, p. 68.

3 *Congressional Record*, 21 September 1970, p. 32902.

4 White, *Air Pollution Emissions from Motor Vehicles*, p. 21.

5 Crandall et al. *Regulating the Automobile*, 1986.

6 Ibid., p. 87.

7 National Commission on Air Quality (NCAQ), *To Breathe Clean Air*, p. 191.

8 In addition, deterioration was blamed for 25 percent, insufficient maintenance for 7 percent, and poor design for 3 percent. These figures are taken from U.S. GAO, *Better Enforcement of Car Emissions Standards*, p. 8.

9 Ibid., p. 11.

10 White, *Air Pollution Emissions from Motor Vehicles*, p. 71; U.S. GAO, *Better Enforcement of Car Emissions Standards*, p. 9.

11 See, for example, the section in *Better Enforcement of Car Emissions Stamdards* entitled "IM Programs are the Most Effective Means for Ensuring that Cars on the Road Meet Emission Standards," p. 9.

12 U.S. GAO, *Vehicle Emissions Inspection and Maintenance Program Is Behind Schedule*, p. 3.

13 The EPA has promulgated specific criteria for programs but has, on various occasions, expressed flexibility in making the states stick to them (ibid., pp. 16–17).

14 U.S. EPA, *Progress in the Prevention and Control of Air Pollution in 1986*, p. IX-6.

15 The Japanese experience with I/M programs provides an interesting contrast. Japanese cars are inspected when bought, then after three years, then every two years, and finally annually when the car is 10 years old or more. Since inspections can cost up to $1000, one

sees few old cars on the road in Japan (General Motors Corporation [GM], *General Motors and the Environment*, interview with Eads, p. 9).

16 Ibid., pp. 27, 21.

17 U.S. EPA, *Progress*, p. IX-3.

18 U.S. GAO, *EPA's Efforts to Control Gasoline Vapors from Motor Vehicles*, p. 2.

19 U.S. GAO, *EPA's Efforts to Control Vehicle Refueling and Evaporation Emissions*, p. 3.

20 The EPA proposal is discussed in depth in U.S. GAO, *EPA's Ozone Policy Is a Positive Step*.

21 The initial Bush proposal is described in *The Washington Post*, 13 June 1989, p. A1.

22 Michael Weisskopf, "House Approves Clean Air Bill," in *The Washington Post*, 24 May 1990, p. A1.

23 NCAQ, *To Breathe Clean Air*, p. 213.

24 U.S. EPA, *Progress*, p. IX-5.

25 U.S. GAO, *Vehicle Emissions*, p. 37.

26 This estimate may be low considering the source is GM, *General Motors and the Environment*, p. 9.

27 EPA data cited in ibid., p. 3.

28 White, *Air Pollution Emissions from Motor Vehicles*, p. 29.

29 *The Washington Post*, 5 June 1989, p. A1.

30 *The Washington Post*, 4 June 1989, p. A19.

31 However, the auto emissions limits on new cars are fairly similar to the New Source Performance Standards provision in stationary source air pollution control.

32 U.S. GAO, *Vehicle Emissions*, p. 9. See also Crandall et al., *Regulating the Automobile*, p. 103.

33 If the latter states did not meet deadline, the EPA could require implementation of I/M programs after that date.

34 Crandall et al., *Regulating the Automobile*, p. 118.

35 U.S. GAO, *Vehicle Emissions*, p. 13.

36 Ibid., pp. 14–16.

37 Ibid., p. 4.

38 Ibid., p. 14.

39 NCAQ, *To Breathe Clean Air*.

40 U.S. GAO, *Vehicle Emissions*, p. 30.

41 U.S. EPA, *Progress*, p. IX-6.

42 GM, *General Motors and the Environment*, p. 4.

43 Ridley, *The State of the States: 1987*, app. 1, p. 3.

44 Jessup, *Guide to State Environmental Programs*, p. 61.

45 Barone and Ujifusa, *Almanac of American Politics 1990*, p. 71.

46 California Air Resources Board (CARB), *California's New Smog Check Program*, p. 1.

47 A Wisconsin administrator told me that California was rated higher than his state overall in air programs only because of their mobile source efforts.

48 From a list provided by CARB dated 13 February 1989.

49 Ridley, *The State of the States: 1987*, app. 1, p. 6.

50 CARB, *California's New Smog Check Program*, p. 1.

51 CARB, *Motor Vehicle Emission Standards Summary*, p. 2.

52 Air Quality Management District (AQMD), *Summary of 1989 Air Quality Management Plan*, p. 16.

53 Ibid., p. 2.

54 U.S. EPA, *Inspection/Maintenance Program Implementation Summary*.

55 CARB, *How to Get a Smog Check*, p. 3.

56 CARB, *California's New Smog Check Program*, p. 4.

57 CARB, *How to Get a Smog Check*, p. 4.

58 Ibid., p. 2.

59 U.S. EPA, *National Air Audit System*, p. 7–4.

60 Ibid., p. 7–7.

61 CARB, *How to Get a Smog Check*, p. 2.

62 Jessup, *Guide to State Environmental Programs*, p. 62.

63 AQMD, *Report This Crime*.

64 AQMD, *Summary of 1989 AQMP*, p. ii; AQMD, *The Challenge of Attainment*, p. iii.

65 Industry spokespersons estimate annual amounts of $15 billion, while others claim the benefits will more than outweigh the total costs. For more, see *Sierra*, July 1989, pp. 16–18.

66 Chilton and Scholtz, *Battling Smog*, p. 20.

67 AQMD, *Summary of 1989 AQMP*, p. 15.

68 Chilton and Scholtz, *Battling Smog*, p. 21.

69 AQMD, *Summary of 1989 AQMP*, p. 17.

70 AQMD, *The Challenge of Attainment*.

71 Frank Clifford, "Air Quality Planners Weigh Commuter Fees," *The Los Angeles Times*, 6 July 1990, p. A1.

72 NCAQ, *To Breathe Clean Air*, p. 197.

73 *The Washington Post*, 6 June 1989, p. A1.

74 CARB, *Motor Vehicle Emission Standards Summary*.

75 Laura Catalano, "Mobile Sources—Major Issues," *New Fuels Report*, 18 June 1990, p. 11.

76 U.S. GAO, *EPA's Efforts to Control Gasoline Vapors from Motor Vehicles*, p. 20.

77 Those states constitute Northeast States for Coordinated Air Use Management and consist of Connecticut, Maine, Massachusetts, New Hampshire, New Jersey, New York, Rhode Island, and Vermont.

78 U.S. GAO, *EPA's Efforts to Control Gasoline Vapors from Motor Vehicles*, p. 19.

79 Ibid., p. 20.

80 Jessup, *Guide to State Environmental Programs*, p. 61.

81 AQMD, *The Challenge of Attainment*, p. 5.

82 *St. Louis Post-Dispatch*, 17 August 1989, p. C6.

83 State and Territorial Air Pollution Program Administrators, *STAPPA's Recommendations for Revising the Clean Air Act*.

84 Davis and Lester (*Federalism amd Environmental Policy*, p. 80) display California as receiving only 24 percent of its FY82 environmental expenditures from the federal government, while the next closest state (New Jersey) receives 36 percent and some states are dependent for as much as 80 percent.

85 Barone and Ujifusa, *Almanac of American Politics 1990*, pp. 69–70.

86 AQMD, *Summary of 1989 AQMP*, p. 12.

87 The overhaul of the Clean Air Act signed into law on 15 November 1990 includes tighter emission controls on automobiles.

5. Nonpoint Source Water Pollution Control

1 Hines, "Legal Aspects of Agriculture's Involvement."

2 The waste treatment total is aggregated from annual totals reported in the U.S. Department of Commerce, *Statistical Abstract*, various years. The total for NPS funding is calculated from unpublished EPA data.

3   Crosson and Brubaker, *Resource and Environmental Effects of U.S. Agriculture*, p. 164.

4   Loehr, *Pollution Control for Agriculture.*

5   U.S. EPA, *Nonpoint Source Guidance.*

6   U.S. EPA, *Unfinished Business.*

7   For a description of these, see Bulletin no. 5 of August 1987 by the USDA, *Conservation/Environmental Protection Programs.*

8   Crosson and Brubaker, *Resource and Environmental Effects of U.S. Agriculture*, p. 10.

9   U.S. Congress, Congressional Research Service, *Agricultural and Environmental Relationships*, p. 178.

10 Zinn and Blodgett, "Agriculture versus the Environment."

11 Crosson, *Conservation Tillage*, p. 32.

12 Public Law 95–217, Sec. 35.

13 U.S. Soil Conservation Service, *National Rural Clean Water Program Manual*, pp. 500–1 to 500–5.

14 Crosson, *Conservation Tillage*, p. 165.

15 H.R. 2100, P.L. 99–198.

16 Congressional Quarterly, *1985 CQA*, pp. 524–39.

17 The 1990 Farm Bill was signed by President Bush on 27 November 1990. The bill contains water quality measures for fragile areas and wetlands, but retains much emphasis on voluntary participation. See Cloud, "1990 Farm Bill"; and Gugliotta, "Farmers, Environmentalists."

18 Kovalic, *The Clean Water Act of 1987*, p. 41.

19 Cook, "Commentary: Agricultural Nonpoint Pollution Control"; Harrington, Krupnick, and Perkin, "Policies for Nonpoint-Source Water Pollution Control."

20 Libby, "Paying the Nonpoint Pollution Control Bill," p. 36.

21 National Research Council, *Groundwater Quality Protection*, p. 23.

22 Crosson, *Conservation Tillage*, p. 187.

23 U.S. GAO, *Greater EPA Leadership.*

24 Henderson et al., *Groundwater Strategies for State Action*, p. 29.

25 U.S. GAO, *Federal-State Environmental Programs*, p. 52.

26 Nagadavera et al. simulated the results of stringent farm controls in Iowa to conclude that "Single state programs tend to place the cost of environmental controls on the farmers of the state where they are applied but bring benefits in income to resource owners elsewhere in the country" (*Soil Conservancy and Environmental Regulations in Iowa*, p. 131).

27 Kramer et al. modeled the effect of differing regulatory controls on farm income in the Chesapeake Bay Area. While the effects varied by the policy (e.g., a 75 percent reduction in pollutants reduced income by 3.2 percent), the study concluded that regulation will "decrease net farm income in the watersheds" and "would be objectionable to many" ("Evaluation of Alternative Policies," p. 845).

28 For a summary, see *The San Francisco Examiner*, 17 March 1985, p. A20.

29 Rosenbaum, *Environmental Politics and Policy*, p. 143.

30 U.S. EPA, *Federal/State/Local Nonpoint Source Task Force*, p. 80.

31 Conservation Foundation, *State of the Environment*, p. 124.

32 Turco and Konopka, *Agricultural Impact on Groundwater Quality.*

33 See also, U.S. EPA, *National Water Quality Inventory* (1987).

34 Crosson, *Conservation Tillage*, p. 165.

35 Crosson and Brubaker, *Resource and Environmental Effects of U.S. Agriculture.*

36 Crosson, *Conservation Tillage*, p. 21.

37 *Congressional Record*, 28 September 1972, p. 32776.

38 The doctrines governing groundwater may differ somewhat from those of surface water. Henderson et al. describe four: (1) absolute ownership where owner has complete freedom to withdraw groundwater, present in Texas and New England; (2) "reasonable use" or the American rule whereby reasonable limits are set on withdrawals and uses; (3) "correlative rights" used in California during droughts where withdrawal is allocated according to proportion of land owned over the groundwater aquifer; and (4) "prior appropriation" where priority is based on overall benefit to the state and the date of application for use (*Groundwater Strategies for State Action*, p. 31).

39 For more on the origins and applications of these contrasting systems of property rights, see Frederick and Hanson, *Water for Western Agriculture*.

40 U.S. EPA, *Nonpoint Source Pollution in the U.S.*

41 U.S. GAO, *The Use of Drinking Water Standards by the States*, p. 28.

42 Ibid.

43 Epp and Shartle, "Commentary: Agricultural Nonpoint Pollution Control," p. 112; Braden and Uchtmann, "Agricultural Nonpoint Pollution Control."

44 Young and Magleby ("Agricultural Pollution Control") reviewed five projects and concluded that two of them (projects in South Dakota and Vermont) enjoyed benefit-cost ratios over 1.

45 Piper et al., "Benefit and Cost Insights," p. 208.

46 Little, "The Economy of Rain," p. 202.

47 Most of this data is provided from a recent analysis by Piper et al., "Benefit and Cost Insights."

48 Young and Magleby, "Agricultural Pollution Control," p. 701.

49 Association of State and Interstate Water Pollution Control Administrators (ASIWPCA), *America's Clean Water: The States' Nonpoint Source Assessment*; ASIWPCA, *America's Clean Water: The States' Nonpoint Source Management Experience.*

50 U.S. EPA, *Nonpoint Source Pollution in the U.S.*, tables b.1 and b.3.

51 Ridley, *The State of the States: 1988.*

52 Crosson and Brubaker, *Resource and Environmental Effects of U.S. Agriculture*, p. 167.

53 Ridley, *The State of the States: 1988*, focus paper 1, p. 6; Ridley, *The State of the States: 1987*, app. 2, p. 3.

54 *The Economist*, 22 August 1987, p. 70.

55 *The Washington Post*, 23 November 1987, p. A3.

56 Wahl, *Promoting Increased Efficiency of Federal Water Use.*

57 Frederick and Hanson, *Water for Western Agriculture*, p. 205.

58 Crosson and Brubaker, *Resource and Environmental Effects of U.S. Agriculture*, p. 167.

59 Iowa Department of Natural Resources (DNR), *State Nonpoint Source Management Report*, p. 3.

60 Iowa DNR, *State Nonpoint Source Assessment Report*, p. 2–15.

61 ASIWPCA, *America's Clean Water: The States' Nonpoint Source Assessment*, Iowa section; Iowa DNR, *State Nonpoint Source Management Report*, p. 5.

62 Crosson and Brubaker, *Resource and Environmental Effects of U.S. Agriculture*, p. 5.

63 Iowa DNR, *State Nonpoint Source Management Report*, p. 6.

64 Ibid., p. 10.

65 Ibid., p. 14.

66 Iowa Cooperative Extension Service, *1989 Progress Report.*

67 Iowa Department of Agriculture, "Ag Calls for Rules for Pesticide Management Areas," *Iowa's Environmental Update* 1, no. 1 (1990): 2.

68 Iowa DNR, *State Nonpoint Source Management Report*, pp. 12–13.

69 Ibid., p. 10.

70 Ibid., p. 8.

71 Ibid., pp. 8–10.

72 Ibid., p. 12.

73 ASIWPCA, *America's Clean Water: The States' Nonpoint Source Management Experience*, pp. IA-1 through IA-3.

74 Piper et al., "Benefit and Cost Insights," pp. 203–8.

75 Statement of James B. Gulliford, Subcommittee on Conservation, Credit, and Rural Development, 9 August 1988.

76 Mike Williams, "SCS Spells Conservation Relief," *The Iowa Farmer Today*, 31 March 1990, p. 1.

77 George Anthan, "USDA Weakens Iowa Soil Rules after Complaint," *The Des Moines Register*, 12 April 1990, p. 1.

78 George Anthan, "Soil Erosion Rules to Get Another Look," *The Des Moines Register*, 21 April 1990, p. 1.

79 See "Take Soil out of Politics" on 22 April 1990 and "A Dirty Deal at the SCS" on 16 May 1990, *The Des Moines Register*.

80 Both the Izaak Walton League and the NRDC opposed the federal actions vociferously.

81 An unpublished letter from the chair of the State Soil Conservation Committee, Oliver Emerson, to the chief of the U.S. SCS on 11 May 1990 termed the action "deplorable" (copy in the possession of the author).

82 Dan Looker, "Angry Iowans Confront Soil Conservation Chief," *The Des Moines Register*, 16 May 1990, p. 1.

83 Padgitt, "Preliminary Results from a Survey of Iowa Soil and Water Commissioners."

84 Looker, "Angry Iowans Confront Soil Conservation Chief."

85 Dan Looker, "Conservation Service Won't Ease Erosion Limits," *The Des Moines Register*, 7 June 1990, p. 1.

86 George Anthan, "USDA Fires SCS Chief Scaling," *The Des Moines Register*, 12 July 1990, p. 55.

87 Unpublished letter from Justin Ward and Thomas Kuhale at NRDC to Scaling, 30 May 1990, copy in possession of the author.

88 Iowa Farm Bureau Federation, *Resolutions*, p. 35.

89 Little, "The Economy of Rain," p. 202.

90 For an interesting description, see Barone and Ujifusa, *Almanac of American Politics 1990*, pp. 425–27.

91 Iowa DNR, *Iowa Groundwater Protection Strategy*, p. 86.

92 U.S. EPA, *Minimization of Hazardous Waste*, p. 1.

93 U.S. EPA, *Nonpoint Source Pollution in the U.S.*, p. 3–1.

6. The State of State Leadership

1 Madison, "Vices of the Political System," p. 364.

2 Davis and Lester, "Federalism and Environmental Policy," p. 62; Harrigan, *Politics and Policy*; Jones, *Clean Air*.

3 Lester, "A New Federalism," p. 59; Van Horn, "The Quiet Revolution," p. 8.

4 Ingram and Mann, "Preserving the Clean Air Act," p. 260; Kraft, Clary, and Tobin, "The Impact of New Federalism"; Lester, "New Federalism and Environmental Policy."

5 Bowman and Kearney, "Dimensions of State Government Capability"; Davis and Lester, "Federalism and Environmental Policy"; Lester, "A New Federalism"; Regens and Reams, "State Strategies for Regulating Groundwater Quality."

6 Freeman, "Air and Water Pollution Policy."

7 Ridley, *The State of the States: 1987*, p. 5.

8 Lester and Keptner, "State Budgetary Commitments"; Scholz and Wei, "Regulatory Enforcement"; Sharkansky, "Government Expenditures and Public Service."

9 Apparently, this has been recognized in the 1990 version of Clean Air legislation.

10 Ingram and Mann, "Interest Groups and Environmental Policy," p. 157.

11 Maidman, "The Greens Party in America."

12 Cohen, "Defusing the Toxic Time Bomb"; Epstein, Brown, and Pope, *Hazardous Waste in America*; Lester and Bowman, "Implementing Environmental Policy"; Williams and Matheny, "Testing Theories of Social Regulation."

13 Epstein, Brown, and Pope, *Hazardous Waste in America*, chap. 9.

14 Vig, "Presidential Leadership," p. 40.

15 Ridley, *The State of the States: 1987*, p. 18.

16 Mazmanian and Morell, "The 'NIMBY' Syndrome," p. 135.

17 Ridley, *The State of the States: 1987*, app. 4.

18 Mazmanian and Morell, "The 'NIMBY' Syndrome," p. 135.

19 Ridley, *The State of the States: 1987*, p. 26.

20 Axelrod, "Energy Policy," pp. 216–18; Ridley, *The State of the States: 1987*, p. 29.

21 Lombard, "Intergovernmental Determinants."

22 Harf and Trout, *The Politics of Global Resources*.

23 Cochran et al., *American Public Policy*, p. 228.

24 Dilger, *National Intergovernmental Programs*, p. 85.

25 Cochran et al., *American Public Policy*, p. 280.

26 Peterson and Rom, "American Federalism, Welfare Policy, and Residential Choices."

27 Tocqueville, *Democracy in America*, p. 169.

Bibliography

Ackerman, Bruce A., and William T. Hassler. 1981. *Clean Coal/Dirty Air*. New Haven: Yale Univ. Press.

Ackerman, Bruce A., Susan Rose-Ackerman, James W. Sawyer, Jr., and Dale W. Henderson. 1974. *The Uncertain Search for Environmental Quality*. New York: The Free Press.

Advisory Commission on Intergovernmental Relations. 1980. *The Federal Role in the Federal System*. Washington, D.C.: ACIR.

——. 1981. *Protecting the Environment: Politics, Pollution, and Federal Policy*. Washington, D.C.: ACIR.

——. 1984. *Regulatory Federalism: Policy, Process, Impact and Reform*. Washington, D.C.: ACIR.

——. 1985. *The Question of State Government Capability*. Washington, D.C.: ACIR.

Aeppel, Timothy. 1987. "Halting Hazardous Waste at the Source." *The Christian Science Monitor*, 30 April, pp. 20–21.

Air Quality Management District. 1989. *Summary of 1989 Air Quality Management Plan*. Los Angeles: AQMD.

——. 1990. *The Challenge of Attainment*. Los Angeles: AQMD.

——. 1990. *Report This Crime*. Los Angeles: AQMD.

Anderson, Frederick R. 1973. *NEPA in the Courts*. Washington, D.C.: Resources for the Future.

Anton, Thomas J. 1984. *Intergovernmental Change in the United States: Myth and Reality*. Ann Arbor: University of Michigan, Institute of Public Policy Studies.

Arnold, R. Douglas. 1979. *Congress and the Bureaucracy*. New Haven: Yale Univ. Press.

Asbell, Bernard. 1978. *The Senate Nobody Knows*. Baltimore: Johns Hopkins Univ. Press.

Ashby, Eric, and Mary Anderson. 1981. *The Politics of Clean Air*. Oxford: Clarendon Press.

Association of State and Interstate Water Pollution Control Administrators. 1984. *America's Clean Water: The States' Evolution of Progress 1972–1982*. Washington, D.C.: GPO.

——. 1985. *America's Clean Water: The States' Nonpoint Source Assessment, 1985*. Washington, D.C.: GPO.

——. 1985. *America's Clean Water: The States' Nonpoint Source Management Experience*. Washington, D.C.: ASIWPCA.

Axelrod, Regina S. 1984. "Energy Policy: Changing the Rules of the Game." In *Environmental Policy in the 1980s*, ed. Norman J. Vig and Michael E. Kraft, pp. 203–25. Washington, D.C.: CQ Press.

Axelrod, Robert. 1981. "The Emergence of Cooperation among Egoists." *American Political Science Review* 75: 306–18.

Bain, Joe S., Richard E. Caves, and Julius Margolis. 1966. *Northern California's Water Industry*. Baltimore: Johns Hopkins Univ. Press.

Ball, Bruce P. 1976. "Water Pollution and Compliance Decision Making." In *Public Policy Making in a Federal System*, ed. Charles O. Jones and Robert D. Thomas, pp. 169–87. Beverly Hills: Sage Publishing Co.

Bardach, Eugene. 1977. *The Implementation Game*. New York: Harper and Row.

Barone, Michael, and Grant Ujifusa. 1990. *The Almanac of American Politics 1990*. Washington, D.C.: National Journal.

Baum, Lawrence. 1989. "State Supreme Courts: Activism and Accountability." In *The State of the States*, ed. Carl E. Van Horn, pp. 103–30. Washington, D.C.: CQ Press.

Baumol, William J., and Wallace E. Oates. 1975. *The Theory of Environmental Policy*. Englewood Cliffs, N.J.: Prentice-Hall.

Beam, David R. 1981. "Washington's Regulation of States and Localities: Origins and Issues." *Intergovernmental Perspectives*, vol. 7, pp. 8–18.

Beam, David R., Timothy J. Conlan, and David B. Walker. 1983. "Federalism: The Challenge of Conflicting Theories and Contemporary Practice." In *Political Science: The State of the Discipline*, ed. Ada W. Finfiter, pp. 247–79. Washington, D.C.: American Political Science Association.

Becker, Stephanie. 1986. *Regulatory Federalism: Policy, Process, Impact and Reform*. Washington, D.C.: Advisory Commission on Intergovernmental Relations.

Beer, Samuel H. 1977. "A Political Scientist's View of Fiscal Federalism." In *The Political Economy of Fiscal Federalism*, ed. Wallace E. Oates, pp. 21–46. Lexington, Mass.: Lexington Books.

Bernstein, Marver H. 1955. *Regulating Business by Independent Commission*. Princeton, N.J.: Princeton Univ. Press.

Beyle, Thad L. 1989. "From Governor to Governors." In *The State of the States*, ed. Carl E. Van Horn, pp. 33–68. Washington, D.C.: CQ Press.

Blomquist, William. 1990. "Exploring State Differences in Groundwater Laws and Policy Innovations, 1980–1989." Paper delivered at the Midwest Political Science Association meetings, Chicago, 5–7 April.

Bluestone, Barry, and Bennett Harrison. 1982. *The Deindustrialization of America*. New York: Basic Books.

Bowman, Ann. 1985. "Hazardous Waste Management: An Emerging Policy Area with an Emerging Federalism." *Publius* 15: 131–44.

Bowman, Ann O'M., and Richard C. Kearney. 1986. *The Resurgence of the States*. Englewood Cliffs, N.J.: Prentice-Hall.

———. 1988. "Dimensions of State Government Capability." *Western Political Quarterly* 41, no. 2: 341–62.

Boudon, Raymond. 1982. *The Unintended Consequences of Social Action*. London: Macmillan.

Boyle, R. H., and R. A. Boyle. 1983. "Acid Rain." *Amicus Journal*, Winter: 22–37.

Braden, John B., and Donald L. Uchtmann. 1985. "Agricultural Nonpoint Pollution Control: An Assessment." *Journal of Soil and Water Conservation* 40, no. 1: 23–26.

Brown, R. Steven, and L. Edward Garner. 1988. *Resource Guide to State Environmental Management*. Lexington, Ky.: The Council of State Governments.

Buresh, James C. 1986. "State and Federal Land Use Regulation: An Application to Groundwater and Nonpoint Source Pollution Control." *Yale Law Journal* 95: 1433–58.

Caldwell, Lynton K. 1975. *Man and His Environment: Policy and Administration*. New York: Harper and Row.

California Air Resources Board. 1990. *California's New Smog Check Program*. Sacramento: CARB.

——. 1990. *How to Get a Smog Check*. Sacramento: CARB.

——. 1990. *Motor Vehicle Emission Standards Summary*. Sacramento: CARB.

Calvert, Jerry W. 1979. "The Social and Ideological Bases of Support for Environmental Legislation: An Examination of Public Attitudes and Legislative Action." *The Western Political Quarterly* 32, no. 3: 327–37.

——. 1989. "Party Politics and Environmental Policy." In *Environmental Politics and Policy*, ed. James P. Lester, pp. 158–78. Durham, N.C.: Duke Univ. Press.

Cheremisinoff, Paul N., and Angelo C. Morresi. 1978. *Air Pollution Sampling and Analysis Deskbook*. Ann Arbor, Mich.: Ann Arbor Science Publishers.

Chernow, Ron. 1978. "The Rabbit that Ate Pennsylvania." *Mother Jones*, January: 18–24.

Chilton, Kenneth, and Anne Scholz. 1989. *Battling Smog: A Plan for Action*. St. Louis: Center for the Study of American Business.

Chubb, John E. 1985. "Federalism and the Bias for Centralization." In *The New Direction in American Politics*, ed. John Chubb and Paul Peterson, pp. 273–306. Washington, D.C.: The Brookings Institution.

——. 1985. "The Political Economy of Federalism." *American Political Science Review* 79, no. 4: 994–1015.

Chubb, John E., and Paul E. Peterson. 1989. "American Political Institutions and the Problem of Governance." In *Can the Government Govern?*, ed. John Chubb and Paul Peterson, pp. 1–43. Washington, D.C.: The Brookings Institution.

Cloud, David S. 1990. "1990 Farm Bill." *Congressional Quarterly Weekly Report*, 1 December: 4037–42.

Cochran, Clarke E., Lawrence C. Mayer, T. R. Carr, and N. Joseph Cayer. 1990. *American Public Policy*. New York: St. Martin's Press.

Cohen, Steven. 1984. "Defusing the Toxic Time Bomb: Federal Hazardous Waste Programs." In *Environmental Policy in the 1980s*, ed. Norman J. Vig and Michael E. Kraft, pp. 273–91. Washington, D.C.: CQ Press.

Cone, Marla. 1989. "Blueprint for Clear Skies," *Sierra*, July/August: 16–18.

Congressional Quarterly. 1978. *Congressional Quarterly Almanac 1977*. Washington, D.C.: CQ Press.

——. 1986. *Congressional Quarterly Almanac 1985*. Washington, D.C.: CQ Press.

Conlan, Timothy. 1988. *New Federalism*. Washington, D.C.: The Brookings Institution.

Conservation Foundation. 1982. *State of the Environment*. Washington, D.C.: Conservation Foundation.

Cook, Ken. 1985. "Commentary: Agricultural Nonpoint Pollution Control: A Time for Sticks?" *Journal of Soil and Water Conservation* 40, no. 1: 105–6.

Copeland, Claudia. 1982. *Federal-State Relations in Transition: Implications for Environmental Policy*. A report for the Library of Congress. Washington, D.C.: GPO.

Corkin, Charles, II. 1978. "Comment on the Political Economy of Implementation." In *Approaches to Controlling Air Pollution*, ed. Ann F. Friedlander, pp. 189–98. Cambridge, Mass.: MIT Press.

Council of State Governments. 1975. *Integration and Coordination of State Environmental Programs*. Lexington, Ky.: Council of State Governments.

——. 1984. *The Book of the States 1984–85*. Lexington, Ky.: Council of State Governments.

Crandall, Robert W. 1981. "Pollution Controls and Publicity Growth in Basic Industries." In

Productivity Measurement in Regulated Industries, ed. Thomas G. Cowing and Rodney E. Stevenson, pp. 347–68. New York: Academic Press.

——. 1983. *Controlling Industrial Pollution*. Washington, D.C.: The Brookings Institution.

Crandall, Robert W., Howard K. Gruenspecht, Theodore E. Keeler, and Lester B. Lave. 1986. *Regulating the Automobile*. Washington, D.C.: The Brookings Institution.

Crenson, Matthew A. 1971. *The Un-Politics of Air Pollution*. Baltimore: Johns Hopkins Univ. Press.

Crosson, Pierre. 1981. *Conservation Tillage and Conventional Tillage: A Comparative Assessment*. Ankeny, Iowa: Soil Conservation Society of America.

Crosson, Pierre R., and Sterling Brubaker. 1982. *Resource and Environmental Effects of U.S. Agriculture*. Washington, D.C.: Resources for the Future.

Croy, James B. 1970. "Federal Supersession: The Road to Domination." *State Government* 48, no. 1: 32–36.

Culhane, Paul J. 1981. *Public Lands Politics*. Baltimore: Johns Hopkins Univ. Press.

Dales, J. H. 1968. *Pollution, Property, and Prices*. Toronto: Univ. of Toronto Press, 1968.

Davies, J. Clarence. 1984. "Environmental Institutions and the Reagan Administration." In *Environmental Policy in the 1980s*, ed. Norman J. Vig and Michael E. Kraft, pp. 143–60. Washington, D.C.: CQ Press.

Davies, J. Clarence, III, and Barbara S. Davies. 1975. *The Politics of Pollution*. Indianapolis: Pegasus.

Davies, J. Clarence, III, and Charles F. Lettow. 1974. "The Impact of Federal Institutional Arrangements." In *Federal Environmental Law*, ed. Erica Dolgin and Thomas Guilbert, pp. 126–90. St. Paul: West Publishing Company.

Davis, Charles E., and James P. Lester. 1987. "Decentralizing Federal Environmental Policy." *Western Political Quarterly* 40, no. 3: 555–66.

——. 1989. "Federalism and Environmental Policy." In *Environmental Politics and Policy*, ed. James P. Lester, pp. 57–84. Durham, N.C.: Duke Univ. Press.

Davis, Martin H., Erhard F. Joeres, and Jeffery J. Peirce. 1980. "Phosphorous Pollution Control in the Lake Michigan Watershed." *Policy Analysis* 6, no. 1: 47–60.

Dawson, Richard E., and James A. Robinson. 1963. "Inter-Party Competition, Economic Variables, and Welfare Policies in the American States." *Journal of Politics* 25: 265–89.

Degler, Stanley E. 1970. *State Air Pollution Control Laws*. Washington, D.C.: Bureau of National Affairs.

Delogu, Orlando E. 1974. *United States Experience with the Preparation and Analysis of Environmental Impact Statements: The National Environmental Policy Act*. Switzerland: International Union for Conservation of Nature and Natural Resources.

Derthick, Martha. 1970. *The Influence of Federal Grants*. Cambridge, Mass.: Harvard Univ. Press.

Derthick, Martha, and Paul J. Quirk. 1985. *The Politics of Deregulation*. Washington, D.C.: The Brookings Institution.

Dilger, Robert Jay. 1989. *National Intergovernmental Programs*. Englewood Cliffs, N.J.: Prentice-Hall.

Downing, Paul B. 1981. "A Political Economy Model of Implementing Pollution Laws." *Journal of Environmental Economics and Management* 8: 255–71.

Downing, Paul B., and Gordon L. Brady. 1979. "Constrained Self-Interest and the Formation of Public Policy." *Public Choice* 34: 15–28.

Dubnick, Mel, and Alan Gitelson. 1981. "Nationalizing State Policies." In *The Nationalization of State Government*, ed. Jerome J. Hanus, pp. 39–74. Lexington, Mass.: Lexington Books.

Duerksen, Christopher J. 1983. *Environmental Regulation of Industrial Plant Siting*. Washington, D.C.: Conservation Foundation.

Dye, Thomas R. 1966. *Politics, Economics, and the Public: Policy Outcomes in the American States*. Chicago: Rand McNally.

Eberts, Randall W., and Timothy J. Grunberg. 1981. "Jurisdictional Homogeneity and the Tiebout Hypothesis." *Journal of Urban Economics* 10: 227–39.

Editors. 1986. "Acid Rain Legislation: A Status Report." *Public Utilities Fortnightly*, 18 September: 34–36.

Editors. 1987. "Clean up, or else." *The Economist*, 22 August: 26.

Edner, Sheldon. 1976. "Intergovernmental Policy Development: The Importance of Problem Definition." In *Public Policy Making in a Federal System*, ed. Charles O. Jones and Robert D. Thomas, pp. 149–67. Beverly Hills: Sage Publications.

Eichbaum, William. 1984. "The Chesapeake Bay: Major Research Program Leads to Innovative Implementation." *Environmental Law Reporter* 6: 10237–10245.

Elazar, Daniel J. 1962. *The American Partnership*. Chicago: Univ. of Chicago Press.

——. 1982. "American Federalism Today: Practice versus Principle." In *American Federalism: A New Partnership for the Republic*, ed. Robert B. Hawkins, Jr., pp. 37–58. San Francisco: Institute for Contemporary Studies.

——. [1966] 1984. *American Federalism: A View from the States*. 3d ed. New York: Harper and Row.

Environmental Law Institute. 1975. *Enforcement of Federal and State Water Pollution Controls*. Washington, D.C.: E.L.I.

——. 1978. *The Response to State and Local Regulations on Emissions to the Atmosphere*. Washington, D.C.: E.L.I.

Epp, Donald J., and James S. Shortle. 1985. "Commentary: Agricultural Nonpoint Pollution Control: Voluntary or Mandatory?" *Journal of Soil and Water Conservation* 40, no. 1: 111–14.

Epstein, Samuel S., Lester O. Brown, and Carl Pope. 1982. *Hazardous Waste in America*. San Francisco: Sierra Club.

Erikson, Robert S., John P. McIver, and Gerald C. Wright, Jr. 1987. "State Political Culture and Public Opinion." *American Political Science Review* 81, no. 3: 797–813.

Erskine, Hazel. 1972. "The Polls: Pollution and Its Costs." *Public Opinion Quarterly* 36, no. 1: 121–22.

Esposito, John C. 1970. "Air Pollution: Moving Beyond Motherhood." In *The Voter's Guide to Environmental Politics*, ed. Garrett deBell, pp. 36–51. New York: Ballantine.

Ferejohn, John A. 1974. *Pork Barrel Politics*. Stanford: Stanford Univ. Press.

Fiorina, Morris P. 1982. "Legislative Choice of Regulatory Forms." *Public Choice* 39: 33–66.

Fiorina, Morris P., and Roger G. Noll. 1978. "Voters, Bureaucrats and Legislators." *Journal of Public Economics* 9: 239–54.

Frederick, Kenneth D., and James C. Hanson. 1982. *Water for Western Agriculture*. Washington, D.C.: Resources for the Future.

Freeman, A. Myrick. 1978. "Air and Water Pollution Policy." In *Current Issues in U.S. Environmental Policy*, ed. Paul R. Portney, pp. 12–67. Baltimore: Johns Hopkins Univ. Press.

Freeman, A. Myrick, and Robert Haveman. 1972. "Clean Rhetoric and Dirty Water." *The Public Interest* 28: 51–65.

Frieden, Bernard J. 1979. *The Environmental Protection Hustle*. Cambridge, Mass.: The MIT Press.

Game, Kingsley W. 1979. "Controlling Air Pollution: Why Some States Try Harder." *Policy Studies Journal* 7, no. 4: 728–38.

Garland, Carol J. 1986. "A Break in the Acid Rain Clouds." *Hamline Law Review* 9: 613–47.

Garnett, James L. 1980. *Reorganizing State Government: The Executive Branch*. Boulder, Col.: Westview Press.

Garvey, D. B., and D. G. Streets. 1980. *In Pursuit of Clean Air: A Data Book of Problems and Strategies at the State Level*. Washington, D.C.: U.S. Department of Energy.

General Motors Corporation. 1989. *General Motors and the Environment*. Detroit, Mich.: GM Corporation.

Goggin, Malcolm L., Ann O'M. Bowman, James P. Lester, and Laurence J. O'Toole, Jr. 1990. *Implementation Theory and Practice*. Glenview, Ill.: Scott, Foresman.

Gormley, William T., Jr. 1989. "Custody Battles in State Administration." In *The State of the States*, ed. Carl E. Van Horn, pp.131–51. Washington, D.C.: CQ Press.

Gramlich, E. 1970. "State and Local Governments and their Budget Constraints." *International Economic Review* 10: 163–81.

Gramlich, Edward M. 1977. "Intergovernmental Grants: A Review of the Empirical Literature." In *The Political Economy of Fiscal Federalism*, ed. Wallace E. Oates, pp. 219–39. Lexington, Mass.: D. C. Heath.

———. 1987. "Federalism and Federal Deficit Reduction." *National Tax Journal* 40, no. 3: 299–313.

Grodzins, Morton. 1960. *Goals for Americans*. Englewood Cliffs, N.J.: Prentice-Hall.

Grunbaum, Werner F., and Lettie M. Wenner. 1980. "Comparing Environmental Litigation in State and Federal Courts." *Publius*, Summer: 129–43.

Gugliotta, Guy. 1990. "Farmers, Environmentalists in Conservation Accord." *The Washington Post*, 15 June, p. A10.

Hahn, Robert W., and Roger G. Noll. 1982. "Designing a Market for Tradable Emissions Permits." In *Reform of Environmental Regulation*, ed. Wesley A. Magat, pp. 119–46. Cambridge, Mass.: Ballinger Publishing Co.

Hamilton, Alexander, James Madison, and John Jay. 1961. *The Federalist Papers*. New York: New American Library.

Hanus, Jerome J. 1981. "Authority Costs in Intergovernmental Relations." In *The Nationalization of State Government*, Jerome J. Hanus, pp. 1–38. Lexington, Mass.: Lexington Books.

Hardin, Garrett. 1977. "The Tragedy of the Commons." In *Managing the Commons*, ed. Garrett Hardin and John Baden, pp. 16–30. San Francisco: W.H. Freeman and Co.

Harf, James E., and B. Thomas Trout. 1986. *The Politics of Global Resources*. Durham, N.C.: Duke Univ. Press.

Harrigan, John. 1980. *Politics and Policy in States and Communities*. Boston: Little, Brown.

Harrington, Winston, and Alan J. Krupnick. 1981. "Stationary Source Pollution Policy and Choices for Reform." In *Environmental Regulation and the U.S. Economy*, ed. Henry M. Peskin, Paul R. Portney, and Allen V. Kneese, pp. 105–30. Baltimore: Johns Hopkins Univ. Press.

Harrington, Winston, Alan J. Krupnick, and Harry M. Peskin. 1985. "Policies for Nonpoint-Source Water Pollution Control." *Journal of Soil and Water Conservation* 40, no. 1: 27–32.

Harris, Richard W., William D. Jeffrey, and Blair W. Steward. 1974. *Interstate Environmental Problems*. Stanford: Stanford Law School.

Hart, Stuart L., and Gordon A. Enk. 1980. *Green Goals and Greenbacks: State-Level Environmental Review Programs and their Associated Costs*. Boulder, Col.: Westview Press.

Haskell, Elizabeth H., and Victoria S. Price. 1973. *State Environmental Management*. New York: Praeger Publishers.

Haveman, Robert H., and Gregory B. Christiansen. 1981. "Environmental Regulations and Productivity Growth." In *Environmental Regulation and the U.S. Economy*, ed. Henry M. Peskin, Paul R. Portney, and Allen V. Kneese, pp. 55–75. Baltimore: Johns Hopkins Univ. Press.

Hawkins, Robert B., Jr. 1982. "American Federalism Again at the Crossroads." In *American Federalism: A New Partnership for the Republic*, ed. Robert B. Hawkins, Jr., pp. 3–15. San Francisco: Institute for Contemporary Studies.

Hawkins, Keith. 1984. *Environment and Enforcement*. Oxford: Clarendon Press.

Heath, Milton S., Jr. 1972. *A Comparative Study of State Water Pollution Control Laws and Programs*. Chapel Hill, N.C.: Water Resources Research Institute.

Hebert, F. Ted, and Deil Wright. 1982. "State Administrators: How Representative? How Professional?" *State Government* 55: 22–28.

Henderson, Timothy R., Jeffrey Trauberman, and Tara Gallagher. 1984. *Groundwater Strategies for State Action*. Washington, D.C.: Environmental Law Institute.

Hines, N. William. 1970. "Legal Aspects of Agriculture's Involvement in Polluted and Clean Water." In *Agricultural Practices and Water Quality*, ed. Ted L. Willrich and George E. Smith, pp. 365–76. Ames, Iowa: Iowa State Univ. Press.

Hirschman, Albert O. 1970. *Exit, Voice, and Loyalty*. Cambridge, Mass.: Harvard Univ. Press.

Holbrook-Provow, Thomas M., and Steven C. Poe. 1987. "Measuring State Political Ideology." *American Politics Quarterly* 15, no. 3: 399–416.

Howitt, Arnold M. 1984. *Managing Federalism*. Washington, D.C.: CQ Press.

Ingram, Helen. 1977. "Policy Implementation Through Bargaining: The Case of Federal Grants-in-Aid." *Public Policy* 25: 499–526.

Ingram, Helen, Nancy Laney, and John R. McCain. 1979. "Water Scarcity and the Politics of Plenty in the Four Corners States." *The Western Political Quarterly* 32, no. 3: 298–306.

Ingram, Helen M. and Dean E. Mann. 1984. "Preserving the Clean Water Act: The Appearance of Environmental Victory." In *Environmental Policy in the 1980s*, ed. Norman J. Vig and Michael E. Kraft, pp. 251–71. Washington, D.C.: CQ Press.

———. 1989. "Interest Groups and Environmental Policy." In *Environmental Politics and Policy*, ed. James P. Lester, pp. 135–57. Durham, N.C.: Duke Univ. Press.

Inman, Robert P. 1979. "The Fiscal Performance of Local Governments: An Interpretative Review." In *Current Issues in Urban Economics*, ed. Peter Mieszkowski and Mahlon Straszheim, pp. 270–321. Baltimore: Johns Hopkins Univ. Press.

———. 1985. "Fiscal Allocations in a Federalist Economy: Understanding the 'New' Federalism." In *American Domestic Priorities*, ed. John M. Quigley and Daniel L. Rubinfeld, pp. 3–33. Berkeley: Univ. of California Press.

Iowa Cooperative Extension Service. 1989. *1989 Progress Report*. Ames, Iowa: Iowa State University.

Iowa Department of Natural Resources. 1987. *Iowa Groundwater Protection Strategy*. Des Moines, Iowa: DNR.

———. 1988. *State Nonpoint Source Assessment Report—Iowa 1988*. Des Moines, Iowa: DNR.

———. 1989. *State Nonpoint Source Management Report*. Des Moines, Iowa: DNR.

———. 1990. *Iowa's Environmental Update*. Des Moines, Iowa: DNR.

Iowa Farm Bureau Federation. 1990. *Resolutions*. Des Moines, Iowa: IFBF.

Jacobs, Jerry. 1979. *Bidding for Business: Corporate Auctions and the 50 Disunited States*. Washington, D.C.: Public Interest Research Group.

Jessup, Deborah Hitchcock. 1988. *Guide to State Environmental Programs.* Washington, D.C.: Bureau of National Affairs.

Jones, Bryan D., Saadia R. Greenberg, Clifford Kaufman, and Joseph Drew. 1977. "Bureaucratic Response to Citizen-Initiated Contacts: Environmental Enforcement in Detroit." *American Political Science Review* 71, no. 1: 148–65.

Jones, Charles O. 1975. *Clean Air.* Pittsburgh, Penn.: Univ. of Pittsburgh Press.

———. 1976. "Regulating the Environment." In *Politics in the American States.* ed. Herbert Jacob and Kenneth N. Vines, pp. 388–427. Boston: Little, Brown.

Kahn, Jeffrey. 1986. "Is Big Sur Really Safe?" *Sierra* 7: 33–41.

Kalt, Joseph P., and Mark A. Zupan. 1984. "Capture and Ideology in the Economic Theory of Politics." *American Economic Review* 74, no. 3: 279–300.

Katzmann, Robert A. 1981. *Regulatory Bureaucracy.* Cambridge, Mass.: MIT Press.

Kaufman, Herbert. 1960. *The Forest Ranger.* Baltimore: Johns Hopkins Univ. Press.

Kay, David, and Harold Jacobson. 1983. *Environmental Protection.* Osmun: Allan Held Publishers.

Keene, John C. 1984. "Innovative Solutions to the Age-Old Problem of Agricultural Pollution." *Zoning and Planning Law Report* 10: 65–71.

Key, V. O. 1949. *Southern Politics.* New York: Vintage Books.

Kieschnick, Michael. 1983. "Taxes and Growth: Business Incentives and Economic Development." In *State Taxation Policy*, ed. Michael Barker, pp. 155–242. Durham, N.C.: Duke Univ. Press.

King, Jonathan F. 1987. "Some States Won't Wait on Acid Rain." *Sierra* 72, no. 3: 18–19.

Kneese, Allen V., and Charles L. Schultze. 1975. *Pollution, Prices and Public Policy.* Washington, D.C.: The Brookings Institution.

Kovalic, Joan M. 1987. *The Clean Water Act of 1987.* 2d ed. Washington, D.C.: Water Pollution Control Federation.

Kraft, Michael E. 1990. "Environmental Gridlock: Searching for Consensus in Congress." In *Environmental Policy in the 1990s*, ed. Norman J. Vig and Michael E. Kraft, pp. 103–24. Washington, D.C.: CQ Press.

Kraft, Michael E., Bruce B. Clary, and Richard J. Tobin. 1988. "The Impact of New Federalism on State Environmental Policy: The Great Lakes States." In *The Midwest Response to the New Federalism*, ed. Peter K. Eisinger and William Gormley, pp. 204–33. Madison: Univ. of Wisconsin Press.

Kramer, R. A., W. T. McSweeney, W. R. Kerns, and R. W. Stavros. 1984. "An Evaluation of Alternative Policies for Controlling Agricultural Nonpoint Source Pollution." *Water Resources Bulletin* 20, no. 6: 841–46.

Lake, Elizabeth E., William M. Hanneman, and Sharon M. Oster. 1977. *Who Pays for Clean Water?* Boulder, Col.: Westview Press.

Lake, Laura. 1982. *Environmental Regulation.* New York: Praeger Publishers.

Lave, Lester B. 1980. "Health, Safety, and Environmental Regulations." In *Setting National Priorities: Agenda for the 1980s*, ed. Joseph A. Pechman, pp. 131–68. Washington, D.C.: The Brookings Institution.

Lehne, Richard. 1972. "Benefits in State-National Relations." *Publius* 2, no. 2: 75–93.

Leone, Robert A., and John C. Jackson. 1981. "The Political Economy of Federal Regulatory Activity: The Case of Water Pollution Controls." In *Studies in Public Regulation*, ed. Gary Fromm, pp. 231–71. Cambridge, Mass.: The MIT Press.

Lester, James P. 1986. "New Federalism and Environmental Policy." *Publius* 16: 149–65.

———. 1990. "A New Federalism: Environmental Policy in the States." In *Environmental Policy in the 1990s*, ed. Norman J. Vig and Michael E. Kraft, pp. 59–79. Washington, D.C.: CQ Press.

Lester, James P., and Ann O'M Bowman. 1989. "Implementing Environmental Policy in a Federal System: A Test of the Sabatier-Mazmanian Model." *Polity* 21, no. 4: 731–53.

Lester, James P., James Franke, Ann Bowman, and Kenneth Kramer. 1983. "Hazardous Wastes, Politics and Public Policy: A Comparative State Analysis." *The Western Political Quarterly* 36: 257–85.

Lester, James P., and Patrick M. Keptner. 1984. "State Budgetary Commitments to Environmental Quality under Austerity." In *Western Public Lands*, ed. John G. Francis and Richard Ganzel, pp. 193–214. Totowa, N.J.: Rowman and Allanheld Publishers.

Lester, James P., and Emmett N. Lombard. 1990. "The Comparative Analysis of State Environmental Policy." *Natural Resources Journal* 30: 301–19.

Libby, Lawrence W. 1985. "Paying the Nonpoint Pollution Control Bill." *Journal of Soil and Water Conservation* 40, no. 1: 33–36.

Lieber, Harvey. 1975. *Federalism and Clean Waters*. Lexington, Mass.: D. C. Heath.

Lind, Douglas. 1981. "Umbrella Equities: Use of the Federal Common Law of Nuisance to Catch the Fall of Acid Rain." *Urban Law Annual* 21: 143–61.

Lindblom, Charles E. 1959. "The Science of 'Muddling Through.'" *Public Administration Review* 19: 79–88.

Liroff, Richard A. 1980. *Air Pollution Offsets*. Washington, D.C.: Conservation Foundation.

Little, Charles E. 1989. "The Economy of Rain and the Tillamook Imperative." *Journal of Soil and Water Conservation* 44, no. 3: 199–202.

Loehr, Raymond C. 1984. *Pollution Control for Agriculture*. 2d ed. New York: Academic Press.

Lombard, Emmett N. 1989. "Intergovernmental Determinants of State Air Quality Policy Outputs: A Comparative Analysis, 1975–87." A paper delivered at the Western Political Science Association meetings, Salt Lake City.

Long, Norton C. 1966. "New Tasks for All Levels of Government." In *Environmental Quality in a Growing Economy*, ed. Henry Jarrett, pp. 141–55. Baltimore: Johns Hopkins Univ. Press.

Lowe, Philip, and Jane Goyder. 1983. *Environmental Groups in Politics*. London: George Allen.

Lowi, Theodore J. 1964. "American Business, Public Policy, Case-Studies, and Political Theory." *World Politics* 16: 667–715.

———. 1969. *The End of Liberalism*. New York: W. W. Norton.

———. 1978. "Europeanization of America? From United States to United State." In *Nationalizing Government: Public Policies in America*, ed. Theodore J. Lowi and Alan Stone, pp. 15–29. Beverly Hills: Sage Publishers.

Luken, Ralph A., and Edward H. Pechan. 1977. *Water Pollution Control*. New York: Praeger Publishers.

Lundqvust, Lennart J. 1980. *The Hare and the Tortoise: Clean Air Policies in the United States and Sweden*. Ann Arbor: Univ. of Michigan Press.

Maass, Arthur. 1959. "Division of Powers: An Areal Analysis." In *Area and Power*, ed. Arthur Maass, pp. 9–26. Glencoe, Ill.: The Free Press.

McConnell, Grant. 1966. *Private Power and American Democracy*. Boston: Houghton Mifflin.

McCubbins, Matthew, and Thomas Schwartz. 1984. "Congressional Oversight Overlooked: Police Patrols versus Fire Alarms." *American Journal of Political Science* 28: 165–79.

Madison, James. [1787] 1901. "Vices of the Political System of the U. States." In *The Writings of James Madison*, vol. 2, 1783–1787, ed. Gaillard Hunt, pp. 361–69. New York: G. P. Putnam's Sons.

Magat, Wesley A., Alan J. Krupnick, and Winston Harrington. 1986. *Rules in the Making.* Washington, D.C.: Resources for the Future.

Magazine, Alan H. 1977. *Environmental Management in Local Government: A Study of Local Response to Federal Mandate.* Westport, Conn.: Greenwood.

Maidman, Randi. 1990. "The Greens Party in America." Unpublished manuscript. St. Louis, Mo.: Washington University.

Manners, Ian R., and Gunders Rudzitis. 1981. "Federal Air Quality Legislation: Implications for Land Use." In *Federalism and Regional Development*, ed. George W. Hoffman, pp. 479–527. Austin: Univ. of Texas Press.

Marcus, Alfred A. 1980. *Promise and Performance.* Westport, Conn.: Greenwood.

Marcus, Melvin G. 1981. "Federal Impacts on Energy Development and Environmental Management in the American West." In *Federalism and Regional Development*, ed. George W. Hoffman, pp. 528–66. Austin: Univ. of Texas Press.

Mashaw, Jerry, and Susan Rose-Ackerman. 1983. "Federalism and Regulation." Unpublished manuscript, Columbia Law School.

Maxwell, James A. 1946. *The Fiscal Impact of Federalism in the United States.* Cambridge, Mass.: Harvard Univ. Press.

May, R. J. 1969. *Federalism and Fiscal Adjustment.* Oxford: Clarendon Press.

Mayhew, David R. 1974. *Congress: The Electoral Connection.* New Haven: Yale Univ. Press.

———. 1986. *Placing Parties in American Politics.* Princeton, N.J.: Princeton Univ. Press.

Mazmanian, Daniel, and David Morell. 1990. "The 'NIMBY' Syndrome: Facility Siting and the Failure of Democratic Discourse." In *Environmental Policy in the 1990s*, ed. Norman J. Vig and Michael E. Kraft, pp. 125–43. Washington, D.C.: CQ Press.

Mazmanian, Daniel A., and Jeanne Nienaber. 1979. *Can Organizations Change?* Washington, D.C.: The Brookings Institution.

Melnick, R. Shep. 1983. *Regulation and the Courts: The Case of the Clean Air Act.* Washington, D.C.: The Brookings Institution.

Miernyk, William H., and John T. Sears. 1974. *Air Pollution Abatement and Regional Economic Development.* Lexington, Mass.: D. C. Heath.

Miller, Doug G. 1981. "Acid Rain: New York Suit Tries to Prove Damage Source in Midwest." *L.A. Daily Journal*, 15 May: 3.

Mitnick, Barry M. 1980. *The Political Economy of Regulation.* New York: Columbia Univ. Press.

Moakley, A. R., and Maureen Moakley. 1984. *The Political Life of the American States.* New York: Praeger.

Moe, Terry M. 1980. *The Organization of Interests.* Chicago: Univ. of Chicago Press.

———. 1985. "Control and Feedback in Economic Regulation: The Case of the NLRB." *American Political Science Review* 79, no. 4: 1094–117.

———. 1989. "The Politics of Bureaucratic Structure." In *Can the Government Govern?*, ed. John E. Chubb and Paul E. Peterson, pp. 267–329. Washington, D.C.: The Broookings Institution.

Montesquieu, Baron de. [1748] 1959. *The Spirit of the Laws.* New York: Hafner Publishing.

Monypenny, Phillip. 1960. "Federal Grants-in-Aid to State Governments: A Political Analysis." *National Tax Journal* 13: 11–16.

Muller, Thomas, and Michael Fix. 1980. "Federal Solicitude, Local Costs: The Impact of Federal Regulation on Municipal Finances." *Regulation* 4, no. 4: 29–36.

Murphy, Earl Finbar. 1977. *Nature, Bureaucracy, and the Rules of Property.* Amsterdam: North-Holland Publishing.

Nagadavera, Vishnuprasad, Earl O. Heady, and Kenneth J. Nicol. 1975. *Implications of Application of Soil Conservancy and Environmental Regulations in Iowa within a National Framework*. Ames, Iowa: Center for Agricultural and Rural Development.

Naisbitt, John. 1984. *Megatrends*. New York: Warner Books.

Nathan, Richard P. 1989. "The Role of the States in American Federalism." In *The State of the States*, ed. Carl E. Van Horn, pp. 15–32. Washington, D.C.: CQ Press.

Nathan, Richard P., Fred C. Doolittle, and Associates. 1985. "The Consequences of Cuts." In *American Intergovernmental Relations*, ed. Laurence J. O'Toole, pp. 260–64. Washington, D.C.: CQ Press.

National Commission on Air Quality. 1981. *To Breathe Clean Air*. Washington, D.C.: NCAQ.

National Research Council. 1986. *Groundwater Quality Protection: State and Local Strategies*. Washington, D.C.: National Academy Press.

National Roundtable of State Waste Reduction Programs. 1989. *Summary of State Waste Reduction Programs*. Raleigh, N.C.: NRSWRP.

Niskanen, William. 1971. *Bureaucracy and Representative Government*. Chicago: Aldine-Atherton.

Noam, Eli. 1982. "The Choice of Governmental Level in Regulation." *Kyklos* 35: 278–91.

Noll, Roger G., and Bruce M. Owen. 1983. *The Political Economy of Deregulation*. Washington, D.C.: American Enterprise Institute.

North Carolina Department of Environment, Health, and Natural Resources. 1982. *Pollution Prevention Pays: Symposium*. Raleigh, N.C.: Pollution Prevention Program.

———. 1990. *Pollution Prevention Tips*. Raleigh, N.C.: Pollution Prevention Program.

North Carolina Division of Environmental Management. 1988. *Water Quality Progress in North Carolina*. Raleigh, N.C.: DEM.

———. 1988–1989. *Administrative Code*. Raleigh, N.C.: DEM.

———. 1989. *Aquatic Toxicity Testing*. Raleigh, N.C.: DEM.

———. 1989. *Aquatic Toxicity Testing: Questions, Answers, and Common Misunderstandings*. Raleigh, N.C.: DEM.

———. 1990. *Organizational Chart*. Raleigh, N.C.: DEM.

Oates, Wallace E. 1972. *Fiscal Federalism*. New York: Harcourt Brace Jovanovich.

———. 1977. "An Economist's Perspective on Fiscal Federalism." In *The Political Economy of Fiscal Federalism*, ed. Wallace E. Oates, pp. 3–20. Lexington, Mass.: Lexington Books.

Olson, Mancur, Jr. 1965. *The Logic of Collective Action*. New York: Schocken Books.

———. 1969. "The Principle of Fiscal Equivalence: The Division of Responsibilities among Different Levels of Government." *American Economic Review* 59: 479–87.

———. 1982. *The Rise and Decline of Nations*. New Haven: Yale Univ. Press.

Opinion Research Corporation. 1971. "Public Opinion on Key Domestic Issues." Mimeograph. Princeton, N.J.: ORC.

Padgitt, Steve, and Paul Lasley. 1990. "Prelminary Results from a Survey of Iowa Soil and Water Commissioners." Unpublished paper for the Iowa University Extension, 15 May.

Palmer, Kenneth T. 1977. *State Politics in the United States*. New York: St. Martin's Press.

Pashigian, B. Peter. 1982. "Environmental Regulation: Whose Self-Interests are Being Regulated?" Unpublished manuscript. University of Chicago.

Patterson, Samuel C. 1976. "American State Legislatures and Public Policy." In *Politics in the American States*, ed. Herbert Jacob and Kenneth N. Vines, pp. 139–95. Boston: Little, Brown.

Peltzman, Sam. 1976. "Toward a More General Theory of Regulation." *Journal of Law and Economics* 19, no. 2: 211–41.

Peterson, Paul E. 1981. *City Limits*. Chicago: Univ. of Chicago Press.

Peterson, Paul E., Barry G. Rabe, and Kenneth K. Wong. 1986. *When Federalism Works.* Washington, D.C.: The Brookings Institution.

Peterson, Paul E., and Mark Rom. 1989. "American Federalism, Welfare Policy, and Residential Choices." *American Political Science Review* 83, no. 3: 711–28.

Peterson, Paul E., and Kenneth K. Wong. 1985. "Toward a Differentiated Theory of Federalism: Education and Housing Policy in the 1980s." In *Research in Urban Policy*, ed. Terry Nichols Clark, pp. 301–24. Greenwich, Conn: Jai Press.

Piper, Steven, C. E. Young, and Richard Magleby. 1989. "Benefit and Cost Insights from the Rural Clean Water Program." *Journal of Soil and Watᵣ Conservation* 44, no. 3: 203–7.

Polsby, Nelson W. 1980. *Community Power and Political Theory.* New Haven: Yale Univ. Press.

Portney, Paul R. 1981. "The Macroeconomic Impacts of Federal Environmental Regulation." In *Environmental Regulation and the U.S. Economy*, ed. Henry M. Peskin, Paul R. Portney, and Allen V. Kneese, pp. 25–54. Baltimore: Johns Hopkins Univ. Press.

Pressman, Jeffrey L. 1975. *Federal Programs and City Politics.* Berkeley: Univ. of California Press.

Pressman, Jeffrey L., and Aaron Wildavsky. 1973. *Implementation.* Berkeley: Univ. of California Press.

Price, Kent A. 1982. "Introduction and Overview." In *Regional Conflict and National Policy*, ed. Kent A. Price, pp. 1–17. Washington, D.C.: Resources for the Future.

Quarles, John. 1976. *Cleaning Up America.* Boston: Houghton Mifflin.

Rabe, Barry G. 1986. *Fragmentation and Integration in State Environmental Management.* Washington, D.C.: Conservation Foundation.

Ranney, Austin. 1976. "Parties in State Politics." In *Politics in the American States*, ed. Herbert Jacob and Kenneth N. Vines, pp. 51–92. Boston: Little, Brown.

Regens, James L., and Margaret A. Reams. 1988. "State Strategies for Regulating Groundwater Quality." *Social Science Quarterly* 69, no. 1: 53–69.

Renshaw, Vernon, Edward A. Trott, Jr., and Howard L. Friedenberg. 1988. "Gross State Product by Industry 1963–1986." *Survey of Current Business*, 19 May: 30–46.

Ridley, Scott. 1987. *The State of the States: 1987.* Washington, D.C.: Fund for Renewable Energy and the Environment.

———. 1988. *The State of the States: 1988.* Washington, D.C.: Fund for Renewable Energy and the Environment.

Riker, William H. 1964. *Federalism: Origin, Operation, Significance.* Boston: Little, Brown.

———. 1964. "Six Books in Search of a Subject or Does Federalism Exist and Does it Matter?" *Comparative Politics* 10: 135–46.

———. 1975. "Federalism." In *Handbook of Political Science*, ed. Fred I. Greenstein and Nelson W. Polsby, pp. 5: 151–59. Reading, Mass.: Addison-Wesley.

Ringquist, Evan J. 1990. "Regulating State Air Quality: Politics and Results." Paper delivered at the Midwest Political Science Association meetings, Chicago, 5–7 April.

Roberts, Marc J., and Susan O. Farrell. 1978. "The Political Economy of Implementation: The Clean Air Act and Stationary Sources." In *Approaches to Controlling Air Pollution*, ed. Ann F. Friedlander, pp. 152–81. Cambridge, Mass.: The MIT Press.

Rogers, Julie Canham, and O. K. Petersen. 1985. "Air Pollution Across State Boundaries." *Michigan Bar Journal* 2: 175–80.

Rohrer, Daniel M. 1970. *The Environment Crisis.* Skokie, Ill.: National Textbook Company.

Rose-Ackerman, Susan. 1977. "Market Models for Water Pollution Control: Their Strengths and Weaknesses." *Public Policy* 25, no. 3: 383–406.

———. 1979. "Market Models of Local Government: Exit, Voting, and the Land Market." *Journal of Urban Economics* 6: 319–37.

——. 1981. "Does Federalism Matter? Political Choice in a Federal Republic." *Journal of Political Economy* 89: 152–65.

——. 1983. "Beyond Tiebout: Modeling the Political Economy of Local Government." In *Local Provision of Public Services: The Tiebout Model after Twenty-five Years*, ed. George Zodrow, pp. 55–83. New York: Academic Press.

Rosenbaum, Nelson. 1981. "Statutory Structure and Policy Implementation: The Case of Wetlands Regulation." In *Effective Policy Implementation*, ed. Daniel Mazmanian and Paul Sabatier, pp. 63–85. Lexington, Mass.: Lexington Books.

Rosenbaum, Walter A. 1973. *The Politics of Environmental Concern*. New York: Praeger Publishers.

——. 1985. *Environmental Politics and Policy*. Washington, D.C.: CQ Press.

Rosenthal, Alan. 1981. *Legislative Life*. New York: Harper and Row.

——. 1989. "The Legislative Institution: Transformed and at Risk." In *The State of the States*, ed. Carl E. Van Horn, pp. 69–101. Washington, D.C.: CQ Press.

Rowland, C. K., and Roger Marz. 1981. "Gresham's Law: The Regulatory Analogy." *Policy Studies Review* 1: 572–80.

Russell, Clifford S., Winston Harrington, and William J. Vaughan. 1986. *Enforcing Pollution Control Laws*. Washington, D.C.: Resources for the Future.

Sabatier, Paul A., and Barbara J. Klosterman. 1981. "A Comparative Analysis of Policy Implementation under Different Statutory Regimes: The San Francisco Bay Conservation and Development Commission, 1965–1972." In *Effective Policy Implementation*, ed. Daniel A. Mazmanian and Paul A. Sabatier, pp. 169–206. Lexington, Mass.: Lexington Books.

Sabatier, Paul A., and Daniel A. Mazmanian. 1983. *Can Regulation Work?* New York: Plenum Press.

Schattschneider, E. E. 1960. *The Semisovereign People*. Hinsdale, Ill.: Dryden Press.

Scheiber, Harry N. 1966. "The Condition of American Federalism: An Historian's View." In *American Intergovernmental Relations*, ed. Laurence J. O'Toole, Jr., pp. 51–57. Washington, D.C.: CQ Press.

Schlesinger, Joseph A. 1971. "The Politics of the Executive." In *Politics in the American States*, ed. Herbert Jacob and Kenneth N. Vines, pp. 210–37. Boston: Little, Brown.

Schmandt, Jurgen, and Hilliard Roderick. 1985. *Acid Rain and Friendly Neighbors: The Policy Dispute between Canada and the United States*. Durham, N.C.: Duke Univ. Press.

Schmenner, Roger W. 1982. *Making Business Location Decisions*. Englewood Cliffs, N.J.: Prentice-Hall.

Schnapf, David. 1982. "State Hazardous Waste Programs under the Federal Resource Conservation and Recovery Act." *Environmental Law* 12: 679–743.

Scholz, John T., and Feng Heng Wei. 1986. "Regulatory Enforcement in a Federalist System." *American Political Science Review* 80, no. 4: 1249–70.

Schubert, Glendon, and Charles Press. 1964. "Measuring Malapportionment." *American Political Science Review* 58, no. 2: 302–27.

Schulman, Paul R. 1975. "Nonincremental Policy Making: Notes Toward an Alternative Paradigm." *American Political Science Review* 69, no. 4: 1354–70.

Scott, W. Richard. 1981. *Organizations*. Englewood Cliffs, N.J.: Prentice-Hall.

Sharkansky, Ira. 1967. "Government Expenditures and Public Services in the American States." *American Political Science Review* 61, no. 4: 1066–77.

Sharkansky, Ira, and Richard I. Hofferbert. 1969. "Dimensions of State Politics, Economics, and Public Policy." *American Political Science Review* 63, no. 3: 867–79.

Shepsle, Kenneth A. 1986. "Institutional Equilibrium and Equilibrium Institutions." In *Political*

Science: The Science of Politics, ed. Herbert F. Weisberg, pp. 51–81. New York:Agathon Press.

Siebert, Horst, Ingo Walter, and Klaus Zimmerman. 1979. *Regional Environmental Policy.* New York: New York Univ. Press.

Simon, Herbert A. 1945. *Administrative Behavior.* New York: The Free Press.

Skok, James E. 1980. "Federal Funds and State Legislatures: Executive-Legislative Conflict in State Government." *Public Administration Review* 40: 561–67.

Skowronek, Stephen. 1982. *Building a New American State: The Expansion of National Administrative Capacities, 1877–1920.* Boston: Cambridge Univ. Press.

Smith, R. Jeffrey. "Acid Rain Bills Reflect Regional Dispute." *Science* 214: 770–71.

Smith, Regan J. R. 1986. "Playing the Acid Rain Game: A State's Remedies." *Environmental Law* 16: 255–317.

Smith, Steven S., and Christopher J. Deering. 1984. *Committees in Congress.* Washington, D.C.: CQ Press.

Stafford, Howard A., Ethel A. Galzerano, and James A. Kelley. 1983. *The Effects of Environmental Regulations on Industrial Location.* Cincinnati: Univ. of Cincinnati Press.

Starr, Mark. 1983. "Periscope." *Newsweek,* 21 February: 21.

State and Territorial Air Pollution Program Administrators. 1985. *STAPPA's Recommendations for Revising the Clean Air Act.* Washington, D.C.: STAPPA.

Stewart, Richard B. 1977. "Pyramids of Sacrifice?" *Yale Law Journal* 86: 1196–275.

———. 1975. "The Reformation of American Administrative Law." *Harvard Law Review* 88, no. 8: 1669–813.

———. 1982. "The Legal Structure of Interstate Resource Conflicts." In *Regional Conflict and National Policy,* ed. Kent A. Price, pp. 87–109. Washington, D.C.: Resources for the Future.

Stigler, George. 1971. "The Theory of Economic Regulation." *Bell Journal of Economics and Management Science* 2: 3–21.

Stoffel, Jennifer. 1980. "Acid Rain Clouds Coal's Future." *State Government News* 6: 3–6.

Strom, Gerald S. 1973. Congressional Policy-Making and the Federal Waste Treatment Construction Grant Program. Ph.D. diss., University of Illinois at Urbana-Champaign.

Sundquist, James L. 1968. *Politics and Policy.* Washington, D.C.: The Brookings Institution.

———. 1986. "American Federalism: Evaluation, Status, and Prospects." Unpublished manuscript. Washington, D.C.: The Brookings Institution.

Tarlock, A. Dan. 1985. "Prevention of Groundwater Contamination." *Zoning and Planning Law Report* 5: 121–27.

Taylor, Serge. 1984. *Making Bureaucracies Think.* Stanford: Stanford Univ. Press.

Thomas, Robert D. 1971. "Federal-Local Cooperation and its Consequences for State-Level Policy Participation: Water Resources in Arizona." *Publius* 1, no. 2: 77–94.

———. 1976. "Intergovernmental Coordination in the Implementation of National Air and Water Pollution Policies." In *Public Policy Making in a Federal System,* ed. Charles O. Jones and Robert D. Thomas, pp. 129–48. Beverly Hills: Sage Publications.

Thompson, Paul B. 1987. *Agricultural Biotechnology and the Rhetoric of Risk: Some Conceptual Issues.* Discussion Paper. Washington, D.C.: Resources for the Future.

Tiebout, Charles M. 1956. "A Pure Theory of Local Expenditures." *Journal of Political Economy* 10: 416–24.

Tobin, Richard J. 1984. "Revising the Clean Air Act: Legislative Failure and Administrative Success." In *Environmental Policy in the 1980s,* ed. Norman J. Vig and Michael E. Kraft, pp. 227–49. Washington, D.C.: CQ Press.

Tocqueville, Alexis de. [1835] 1945. *Democracy in America*. New York: Random House.

Tolley, George S., Philip E. Graves, and Glenn C. Blomquist. 1981. "Toward Improved Environmental Policy." In *Environmental Policy*, ed. George S. Tolley, pp. 1: 175–96. Cambridge, Mass.: Ballinger Publishing Company.

Tran, Dang T. 1986. "Locational Factors in the Declining Industrial Competitive Advantage of the New York Urban Region." *Journal of Regional Science* 26, no. 1: 121–39.

Truman, David B. 1951. *The Governmental Process*. New York: Alfred A. Knopf.

Turco, Ronald F., and Allan E. Konopka. 1988. *Agricultural Impact on Groundwater Quality*. West Lafayette, Ind.: Water Resources Research Center.

U.S. Bureau of the Census. 1970–80. *Environmental Quality Control*. Washington, D.C.: GPO.

———. 1972. *Census of Manufacturers*. Washington, D.C.: GPO.

———. 1972. *Compendium of Public Employment*. Washington, D.C.: GPO.

———. 1976. *Pollution Abatement Costs and Expenditures*. Washington, D.C.: GPO.

———. 1977. *Census of Manufacturers*. Washington, D.C.: GPO.

———. 1977. *Compendium of Public Employment*. Washington, D.C.: GPO.

U.S. Congress. 1977. *U.S. Code Congressional and Administrative News*. St. Paul, Minn.: West Publishing.

U.S. Congress, Congressional Research Service. 1979. *Agricultural and Environmental Relationships: Issues and Priorities*. Washington, D.C.: GPO.

U.S. Congress, House Committee on Public Works. 1971. *Hearings on Water Pollution Control Legislation*. Washington, D.C.: GPO.

———. 1978. *Hearings on Oversight of Water Pollution Legislation*. Washington, D.C.: GPO.

U.S. Congress, Senate Committee on Environment and Public Works. 1981. *Hearings on Acid Rain*. Washington, D.C.: GPO.

U.S. Congress, Senate Committee on Public Welfare. 1971. *Hearings on Water Pollution Control Legislation*. Washington, D.C.: GPO.

U.S. Congress, Senate Committee on Public Works. 1970. *Hearings on Air Pollution*. Washington, D.C.: GPO.

U.S. Congress, Senate Subcommittee on Environmental Pollution. 1979. *Enforcement of Environmental Regulations*. Washington, D.C.: G.P.O.

U.S. Congressional Budget Office. 1985. *Environmental Regulation and Economic Efficiency*. Washington, D.C.: GPO.

U.S. Council on Environmental Quality. 1973. *Environmental Quality*. Washington, D.C.: GPO.

———. 1980. *Public Opinion on Environmental Quality*. Washington, D.C.: GPO.

———. 1981. *Environmental Quality*. Washington, D.C.: GPO.

———. 1983. *Environmental Quality*. Washington, D.C.: GPO.

U.S. Department of Agriculture. 1987. *Conservation/Environmental Protection Programs*. Washington: USDA.

U.S. Department of Commerce. Various years. *Statistical Abstract of the United States*. Washington, D.C.: U.S. Bureau of the Census.

U.S. Environmental Protection Agency. 1974. *Monitoring and Air Quality Trends Report*. Washington, D.C.: GPO.

———. 1974. *National Water Quality Inventory*. Washington, D.C.: GPO.

———. 1974. *1974 National Emissions Report*. Washington, D.C., GPO.

———. 1976. *Compliance Status of Major Air Pollution Facilities*. Washington, D.C.: GPO.

———. 1977. *A Review of Vehicle Inspection and Maintenance Programs in the United States*. Washington, D.C.: GPO.

——. 1977. *1977 National Emissions Trends Reports*. Washington, D.C.: GPO.

——. 1978. *Compilation and Analysis of State Regulations for SO2, NOx, Opacity, Continuous Monitoring, and Applicable Test Methods*. Washington, D.C.: GPO.

——. 1978. *Interim National Municipal Policy and Strategy*. Washington, D.C.: GPO.

——. 1978. *1974 National Emissions Report*. Research Triangle Park, N.C.: GPO.

——. 1980. *Compliance Data System Users Guide*. Washington, D.C.: GPO.

——. 1980. *National Air Pollutant Emission Estimates, 1940–1980*. Washington, D.C.: GPO.

——. 1980. *Water Quality Management Directory*. Washington, D.C.: GPO.

——. 1984. *Nonpoint Source Pollution in the U.S.* Washington, D.C.: GPO.

——. 1985. *Federal/State/Local Nonpoint Source Task Force and Recommended National Nonpoint Source Policy*. Washington, D.C.: GPO.

——. 1985. *National Air Audit System: FY 1985 National Report*. Washington, D.C.: GPO.

——. 1985. "Privatization of Municipal Wastewater Treatment." In *American Intergovernmental Relations*, ed. Laurence J. O'Toole, pp. 296–301. Washington, D.C.: CQ Press.

——. 1986. *Minimization of Hazardous Waste*. Washington, D.C.: GPO.

——. 1987. *National Water Quality Inventory*. Washington, D.C.: GPO.

——. 1987. *Nonpoint Source Guidance*. Washington, D.C.: GPO.

——. 1987. *Waste Minimization*. Washington, D.C.: GPO.

——. 1988. *Progress in the Prevention and Control of Air Pollution in 1986*. Washington, D.C.: GPO.

——. 1988. *Unfinished Business*. Washington, D.C.: GPO.

——. 1989. *Inspection/Maintenance Program Implementation Summary*. Ann Arbor, Mich.: Office of Mobile Sources.

——. 1989. *National Air Quality and Emissions Trends Reports, 1987*. Washington, D.C.: GPO.

U.S. Environmental Protection Agency and U.S. Fish and Wildlife Service. 1984. *1982 National Fisheries Survey*. Washington, D.C.: GPO.

U.S. General Accounting Office. 1978. *16 Air and Water Pollution Issues Facing the Nation*. Washington, D.C.: GPO.

——. 1979. *Better Enforcement of Car Emission Standards: A Way to Improve Air Quality*. Washington, D.C.: GPO.

——. 1979. *Improvements Needed in Controlling Major Air Pollution Sources*. Washington, D.C.: GPO.

——. 1980. *Costly Wastewater Treatment Plants Fail to Perform as Expected*. Washington, D.C.: GPO.

——. 1980. *Federal-State Environmental Programs—The State Perspective*. Washington, D.C.: GPO.

——. 1984. *EPA Needs to Improve Its Oversight of Air Pollution Control Grant Expenditures*. Washington, D.C.: GPO.

——. 1985. *Air Pollution: Environmental Protection Agency's Inspections of Stationary Sources*. Washington, D.C.: GPO.

——. 1985. *EPA's Delegation of Responsibilities to Prevent Significant Deterioration of Air Quality*. Washington, D.C.: GPO.

——. 1985. *Vehicle Emissions Inspection and Maintenance Program Is Behind Schedule*. Washington, D.C.: GPO.

——. 1986. *EPA Response to Questions on Its Inspection and Maintenance Program*. Washington, D.C.: GPO.

——. 1986. *EPA's Standard Setting Process Should Be More Timely and Better Planned.* Washington, D.C.: GPO.

——. 1986. *Surface Mining: Interior Department and States Could Improve Inspection Programs.* Washington, D.C.: GPO.

——. 1987. *Acid Rain.* Washington, D.C.: GPO.

——. 1987. *EPA's Efforts to Control Vehicle Refueling and Evaporative Emissions.* Washington, D.C.: GPO.

——. 1987. *States Accorded a Major Role in EPA's Air Toxics Strategy.* Washington, D.C.: GPO.

——. 1987. *States Assigned a Major Role in EPA's Air Toxics Strategy.* Washington, D.C.: GPO.

——. 1987. *Surface Mining: States Not Assessing and Collecting Monetary Penalties.* Washington, D.C.: GPO.

——. 1988. *EPA's Efforts to Develop a New Model for Regulating Utility Emissions.* Washington, D.C.:GPO.

——. 1988. *EPA's Ozone Policy Is a Positive Step but Needs More Legal Authority.* Washington, D.C.: GPO.

——. 1988. *Stronger Enforcement Needed to Improve Compliance at Federal Facilities.* Washington, D.C.: GPO.

——. 1988. *The Use of Drinking Water Standards by the States.* Washington, D.C.: GPO.

——. 1989. *EPA's Efforts to Control Gasoline Vapors from Motor Vehicles.* Washington, D.C.: GPO.

——. 1989. *More EPA Action Needed to Improve the Quality of Heavily Polluted Waters.* Washington, D.C.: GPO.

——. 1990. *Greater EPA Leadership Needed to Reduce Nonpoint Source Pollution.* Washington, D.C.: GPO.

——. 1990. *Serious Problems Confront Emerging Municipal Sludge Management Program.* Washington, D.C.: GPO.

U.S. Geological Survey. 1984. *National Water Summary 1983: Hydrologic Events and Issues.* Washington, D.C.: GPO.

U.S. Office of Technology Assessment. 1987. *Serious Reduction of Hazardous Waste.* Washington, D.C.: GPO.

U.S. Soil Conservation Service. 1979. *National Rural Clean Water Program Manual.* Washington, D.C.: GPO.

Van Horn, Carl E., ed. 1989. "The Entrepreneurial States." In *The State of the States*, ed. Carl E. Van Horn, pp. 209–21. Washington, D.C.: CQ Press.

Van Horn, Carl E., ed. 1989. "The Quiet Revolution." In *The State of the States*, ed. Carl E. Van Horn, pp. 1–13. Washington, D.C.: CQ Press.

Vaughan, William J., and Clifford S. Russell. 1982. *Freshwater Recreational Fishing.* Washington, D.C.: Resources for the Future.

Vestigo, James R. 1985. "Acid Rain and Tall Stack Regulation Under the Clean Air Act." *Environmental Law* 15: 711–44.

Vig, Norman J. 1990. "Presidential Leadership: From the Reagan to the Bush Administration." In *Environmental Policy in the 1990s*, ed. Norman J. Vig and Michael E. Kraft, pp. 33–58. Washington, D.C.: CQ Press.

Vig, Norman J., and Kraft, Michael E. 1984. "Environmental Policy from the Seventies to the Eighties." In *Environmental Policy in the 1980s*, ed. Norman J. Vig and Michael E. Kraft, pp. 3–26. Washington, D.C.: CQ Press.

Wahl, Richard W. 1987. *Promoting Increased Efficiency of Federal Water Use Through Voluntary Water Transfers*. Discussion Paper. Washington, D.C.: Resources for the Future.

Walker, David B. 1981. *Toward a Functioning Federalism*. Cambridge, Mass.: Winthrop Publishers.

Walker, Jack L. 1969. "The Diffusion of Innovation among the American States." *American Political Science Review* 63, no. 3: 880–99.

Walker, Richard, and Michael Storper. 1978. "Erosion of the Clean Air Act of 1970: A Study in the Failure of Government Regulation and Planning." *Boston College Environmental Affairs Law Review* 7: 189–257.

Webber, David J. 1985. "Equitably Reducing Transboundary Causes of Acid Rain." In *The Acid Rain Debate*, ed. Ernest J. Yanarella and Randal H. Ihara, pp. 219–38. Boulder, Colo.: Westview Press.

Wenner, Lettie. 1972. "Enforcement of Water Pollution Control Law." *Law and Society Review*, May: 481–507.

——. 1976. *One Environment under Law*. Pacific Palisades, Calif.: Goodyear Publishing.

——. 1982. *The Environmental Decade in Court*. Bloomington: Indiana Univ. Press.

Whitaker, John C. 1976. *Striking a Balance*. Washington, D.C.: American Enterprise Institute.

White, Lawrence J. 1982. *The Regulation of Air Pollutant Emissions from Motor Vehicles*. Washington, D.C.: American Enterprise Institute.

Wiessler, David A. 1982. "As States Fight Over Water, Energy, Jobs." *U.S. News and World Report*, 23 August: 49–50.

Williams, Bruce A. 1983. "Bounding Behavior: Economic Regulation in the American States." In *Politics in the American States*, 4th ed., ed. Virginia Gray, pp. 329–70. Boston: Little, Brown.

Williams, Bruce A., and Albert R. Matheny. 1984. "Testing Theories of Social Regulation: Hazardous Waste Regulation in the American States." *Journal of Politics* 46, no. 1: 428–58.

Wilson, James Q. 1974. "The Politics of Regulation." In *Social Responsibility and the Business Predicament*, ed. James W. McKie, pp. 135–68. Washington, D.C.: The Brookings Institution.

——. 1983. *American Government*. Lexington, Mass.: D. C. Heath.

Wilson, Len U. 1980. "Environmental Quality up to States." *State Government News*, 6: 5–6.

Wisconsin Department of Natural Resources. 1986–89. *Wisconsin Administrative Code*. Madison: DNR.

——. 1988. *Expanding Industry in Wisconsin: A Guide to Meeting Air Quality Requirements*. Madison: DNR.

Wood, B. Dan. 1986. Policy Interventions in Federal Policy Systems. Ph.D. diss., University of Houston.

Young, C. E., and R. S. Magleby. 1987. "Agricultural Pollution Control: Implications from the Rural Clean Water Program." *Water Resources Bulletin* 23, no. 4: 701–7.

Young, Oran. 1981. *Natural Resources and the State*. Berkeley: Univ. of California Press.

Zedrosser, Joseph J. 1985. "Environmental Law." *Syracuse Law Review* 36: 231–35.

Ziegler, L. Herman, and Hendrick Van Dalen. 1976. "Interest Groups in State Politics." In *Politics in the American States*, 2d ed., ed. Herbert Jacob and Kenneth N. Vines, pp. 93–136. Boston: Little, Brown.

Zinn, Jeffrey A., and John E. Blodgett. 1989. "Agriculture versus the Environment: Communicating Perspectives." *Journal of Soil and Water Conservation* 44, no. 3: 184–87.

Zwick, David, and Mary Benstock. 1971. *Water Wasteland*. New York: Grossman Publishers.

Index

William R. Lowry is Associate Professor of Political Science
at Washington University in St. Louis.

Library of Congress Cataloging-in-Publication Data
Lowry, William R.
The dimensions of federalism : state governments and
pollution control policies / William R. Lowry.
Includes bibliographical references and index.
ISBN 0-8223-1162-3 (cl: alk. paper). —
ISBN 0-8223-1819-9 (pa.: alk. paper)
1. Pollution — Government policy — United States — States.
2. Federal government — United States.   I. Title.
HC110.P55L68   1991
363.7'06'0973 — dc20   91-13582 CIP